LANDSCAPES OF GLOBALIZATION

In this critical and sophisticated analysis, Philip F. Kelly challenges the conventional definition of globalization as an irresistible and inevitable force to which societies must succumb. By tracing the consequences of global economic integration in the Philippines, he argues that 'global' processes are constituted, accommodated, mediated and resisted in social processes at multiple scales, from the national economy to the village and the household.

The study harnesses arguments on a broad theoretical plane to new source material derived from close investigation of Manila's rural, but rapidly industrializing hinterland. It therefore promises to inform debate both as an up-to-date assessment of Philippine development in the 1990s, and as an essential contribution to the study of geographies and economies of space.

Philip F. Kelly is Assistant Professor of Southeast Asian Studies at the National University of Singapore.

ROUTLEDGE PACIFIC RIM GEOGRAPHIES
Series Editors: John Connell, Lily Kong and John Lea

1 LANDSCAPES OF GLOBALIZATION
Human geographies of economic change in the Philippines
Philip F. Kelly

LANDSCAPES OF GLOBALIZATION

Human geographies of economic change in the Philippines

Philip F. Kelly

London and New York

First published 2000 by Routledge
11 New Fetter Lane, London EC4P 4EE

Simultaneously published in the USA and Canada
by Routledge
29 West 35th Street, New York, NY 10001

Routledge is an imprint of the Taylor & Francis Group

© 2000 Philip F. Kelly

Typeset in Garamond by Steven Gardiner Ltd., Cambridge
Printed and bound in Great Britain by TJ International Ltd, Padstow, Cornwall

All rights reserved. No part of this book may be reprinted or reproduced
or utilized in any form or by any electronic, mechanical, or other means,
now known or hereafter invented, including photocopying and recording, or
in any information storage or retrieval system, without permission in writing
from the publishers.

British Library Cataloguing in Publication Data
A catalogue record for this book is available from the British Library

Library of Congress Cataloguing in Publication Data
Kelly, Philip F., 1970–
Landscapes of globalization: human geographies of economic change
in the Philippines / Philip F. Kelly.
p. cm.
Includes bibliographical references and index.
1. Philippines – Economic conditions – 1986– I. Title.
HC455.K45 2000   99-36241
330.9599′048–dc21   CIP

ISBN 0-415-19159-9

FOR HAYLEY

CONTENTS

List of illustrations viii
List of tables x
Series preface xii
Preface xiv
Acknowledgements xvi

1 Introduction: putting globalization in its place 1

2 The Philippines and the global economy: constructing a place in the world 16

3 Globalization and the Philippine space economy: patterns, processes and politics 47

4 The globalizing village I: landscapes of labour 73

5 The globalizing village II: the physical landscape 114

6 Resisting and reimagining the global 141

7 Conclusion 159

Appendix: Methodological issues 166

Notes 169
References 175
Index 185

ILLUSTRATIONS

Plates

1	Postcolonial complexity: all-American dolls dressed in Spanish costumes on sale in 1998 to celebrate the centenary of the Philippine revolution against colonial rule	27
2	A company manufacturing garments at the Cavite Export Processing Zone	53
3	Billboards proclaiming Cavite's transformation	75
4	The house of an agricultural labourer on the main street through barrio Bunga	81
5	Planting rice; but housing developments are encroaching	85
6	The CEPZ: a banner entices factory workers to the even richer rewards of the global economy by working overseas	99
7	A new factory development in the midst of prime agricultural land	116
8	A residential subdivision in barrio Mulawin: most plots remain undeveloped as supply outstrips demand	116
9	Rice land now lies idle, awaiting development	117
10	Refuse from a new residential subdivision clogging an irrigation channel	136
11	The campaign that toppled the Kingpin: the Velasco–Revilla ticket, with tacit support from 'FVR', ended Remulla's rule over Cavite	154

Figures

3.1	Regional share of manufacturing value added, 1981–96 (%)	51
3.2	Foreign direct investment to the Philippines, 1985–97 (US$ millions)	51
3.3	Employment in Philippine economic zones, 1986–98	55
4.1	Age and sex structure of Bunga, 1995	95
4.2	Occupational structure of working population in Bunga, 1995	98
5.1	Age and sex structure of Mulawin, 1995	122
5.2	Occupational structure of working population in Mulawin, 1995	125

LIST OF ILLUSTRATIONS

Maps

1	The Philippines	50
2	The national core region and Calabarzon area	59
3	Province of Cavite	62
4	Sketch map of barrio Bunga, June 1995	80
5	Sketch map of barrio Mulawin, June 1995	120

TABLES

3.1	Regional share of GDP, 1981–96	48
3.2	Investment in economic zones by sector, 1992–7	53
3.3	Sources of foreign investment approved by PEZA and BOI	54
3.4	Employment structure of CEPZ enterprises, April 1995	56
3.5	Regional employment generation by BOI projects, 1985–99	57
3.6	Distribution of employment in Philippine economic zones, 1998	58
3.7	Employment generation by BOI-registered investments, 1985–98 in southern Tagalog provinces	60
3.8	Characteristics of Cavite's industrial sector, June 1995	61
3.9	Regional distribution of public investment, 1989–92 (%)	66
4.1	The cropping cycle in Bunga	83
4.2	Rainy season cash expense estimates for a 1-hectare rice farm, 1995	87
4.3	Rainy season rice account estimates for a 1-hectare farm, 1995 (in *cavans*)	88
4.4	Sources of agricultural financing for farmers in Bunga, 1995	89
4.5	Distribution of household chores in Bunga by gender, 1995	92
4.6	Primary and secondary household occupation by gender, Bunga, 1995	96
4.7	Total occupational structure of Bunga, by sector, 1995	97
4.8	Currently deployed overseas workers from Bunga, by destination, 1995	98
4.9	Currently deployed overseas workers from Bunga, by occupation, 1995	98
4.10	The emergence of *sabog* utilization in Bunga, 1975–95	106
4.11	Birthplace and arrival date of migrants in Bunga, 1995	108
4.12	Birthplace and occupation of migrants in Bunga, 1995	108
5.1	Land use in Cavite, 1988	118
5.2	Land conversions in Cavite approved or being processed by DAR, 1988–95	118
5.3	Primary and secondary household occupation by gender, Mulawin, 1995	123
5.4	Independent residents' occupations, Mulawin, 1995	124

LIST OF TABLES

5.5	Total occupational structure of Mulawin, by sector, 1995	124
5.6	Birthplace and arrival date of non-Cavitenos in Mulawin, 1995	126
5.7	Birthplace and occupation of non-Cavitenos in Mulawin, 1995	126
5.8	Overseas workers, by occupation and gender, Mulawin, 1995	127
5.9	Overseas workers, by workplace and gender, Mulawin, 1995	127
5.10	Residents of Monteverde subdivision, by birthplace, 1995	131
5.11	Residents of Monteverde subdivision by occupation, 1995	131

SERIES PREFACE

The Pacific Rim is the world's most dynamic region, confronting diverse economic, political and socio-cultural changes of different orders of magnitude. The region has been open to, and the impetus behind, forces of globalization. The transcending of borders has taken place in political, economic and socio-cultural spheres. Economically, transnational corporations have emerged as strong contenders in the world of capital, establishing and maintaining business networks in varied parts of the world, not least Asian business networks. Culturally, globalizing forces have given rise to transculturation, the interchange of cultural elements and the breaking down of cultural borders. Politically, states have had to be concerned about the maintenance of the 'nation' in the midst of larger cross-national forces. Yet, at the same time, indigenous practices have been asserted through patterns of localization, with local (often ethnic) identities becoming increasingly important as geographical phenomena.

The Pacific Rim is actor and driver in all these processes, as globalization and localization are constantly being reconstructed in new and imaginative ways, and as modernity is transformed and differently imagined. This series seeks to provide a platform for the analysis of such developments in Asia (or parts of Asia), as well as island states at the centre and on the eastern shores of the Pacific Ocean region. It will include scrutiny of issues such as

- environmental change (e.g. resource management, sustainable urban development, deforestation and global warming)
- urban transformations (e.g. city cultures, urban management and planning, migration, housing, global cities, the informal city, tourism and urban change)
- culture and identity (e.g. religion, music, food, fashion, gender)
- local and regional change (e.g. globalization and local autonomy, entrepreneurs, transnational corporations, the informal sector, rural development).

All these, and more, form the empirical situations within which the global and local intersect, and relate to the core themes of culture, environment, urbanization and geographical change that this series seeks to address.

The series is oriented towards specialized tertiary classes focusing on the Asia-

SERIES PREFACE

Pacific region and will appeal to a multidisciplinary audience both despite and because of the geographical derivation. We are happy that Phil Kelly's exciting book on urban transformations in the urban fringes of Manila opens this series. It will be a forerunner of others, dealing, for example, with issues of sexuality and sex workers in Asia, and migration in China. We hope the series will be appealing to many across disciplines.

John Connell
Department of Geography, University of Sydney
connelljohn@hotmail.com

Lily Kong
Department of Geography, National University of Singapore
lilykong@nus.edu.sg

John Lea
Department or Urban Studies, University of Sydney
lea@arch.usyd.edu.au

PREFACE

This is a study motivated, above all, by an interest in how broad processes of social and economic change are experienced in people's lives. It is an attempt to trace the meaning of globalization in a site where, as both a process and an idea, it is having profound impacts. This site is not, however, a major financial centre or some cosmopolitan enclave in a world city. Instead, it is in the landscape of Manila's rural, but industrializing, periphery. It is a place where, in the 1990s in particular, new urban spaces of everyday life are being constructed at a hectic pace. Residential estates are rising on rice fields; industrial zones punctuate the landscape; new migrants are settling in villages that could previously be accounted for with perhaps half a dozen surnames; some 'locals' are leaving to work overseas while others are taking jobs in nearby factories. Across the landscape, billboards herald new developments, and once-productive agricultural land lies in varying states of conversion into other uses. This is not a 'frontier' of capitalism, but it is a place of contemporary developmental intensity in a changing capitalist global economy.

The process driving these dynamic social and physical landscapes is primarily the influx of global capital – foreign direct investment in export-oriented manufacturing industries – arriving as part of yet another iteration in the constant restructuring of the world economy. This is a place whose time has come in the ongoing reworking of spatial divisions of labour on a global scale. It is an emergent node in the global space of flows.

But such interpretations provide an easy analysis of these changes: as the imprint of the global economy on a particular place. At best this is only part of the story. I believe that an account of these transforming landscapes is not just about what globalization means for a small corner of the Philippines. There are also important points to be made concerning what this particular place implies about globalization. Globalization has frequently been represented as an irresistible, inevitable and inexorable force to which places must succumb. The arguments to be developed in this book, however, suggest that a far more sophisticated understanding of contemporary change in the global economy is needed: one that is sensitive to the 'embeddedness' of social processes at multiple scales, and aware of the political power relations implied in understanding globalization as a reified external force. This study, then, is as much about how we understand (particularly economic) globalization, as it is about how globalization is manifested in the Philippines, although it is through a careful examination of such manifestations that broader conclusions are drawn.

This book has no pretensions to being an economic anthropology of the Philippines or even a small part of the country. It does not claim to have discovered any new or essential features of Philippine society or to speak on behalf of 'a people'. In an era when 'representational' scholarship is rightly seen as problematic and Filipino scholars are writing with vigour about their own society, such goals are hardly tenable. The days of entering 'the field' with a missionary zeal to improve the world or unearth the exotic are fortunately largely over. At the same time, however, research beyond one's immediate context (of place, class, culture and identity) should not be 'off-limits'. Instead, I would suggest that an important task facing social researchers, and geographers in particular, is to understand the global processes in which we are all complicit – especially those of us in positions of social, economic or political power – and how they are implicated in transforming the lives of others. For while I will argue that globalization is a political discourse, a myth of sorts, it is also a sufficiently real and material process that the lives of 'distant strangers' are inevitably but unconsciously touched by decisions made elsewhere and processes constituted at larger scales.

What this book attempts to do, then, is to construct an account of how globalization is embedded in multiple scales of social relations in the Philippines – national, regional, provincial, village, household, individual. In particular, it examines the spatial and social traces of the global economy in the Philippines, both in a historical context and in contemporary transformations of social and physical landscapes on Manila's periphery. But in addition to portraying the changes wrought in a zone of intensive globalized development, this analysis also has implications for the way in which we think about the global scale. Rather than being a determinative and disembodied force, the global becomes instead a set of processes constituted simultaneously in multiple, nested, places of varying scales. At each scale, 'global' processes are received, interpreted, constructed, accommodated, mediated and resisted in different ways and in a manner that generally serves to legitimize and entrench existing power relations. Moreover, where changes resulting from integration with the global economy are challenged, this resistance may also operate on multiple scales, including the global.

This study, then, ranges across scales, always treating globalization as a set of simultaneously material and discursive processes embedded in a political economy of power relations. The arguments developed here have implications for the ways in which development is understood and experienced in the Philippines and issues are raised regarding how the country relates to the space of the global economy. But I hope that, as globalized development becomes ever more widely represented as the only path to prosperity, and more and more places come under its thrall, the type of approach adopted here might also point to a style of analysis that encompasses the complexity of contemporary economic life and addresses the resulting changes in a sophisticated and critical manner.

ACKNOWLEDGEMENTS

In the Philippines, my first debt of gratitude is to my *lola*, Mrs Rufina Solis of Tanza, who adopted me as her *apo* (grandson) for the duration of my fieldwork and treated me with commensurate kindness and hospitality. My life in Tanza, and my research, were also greatly enriched by Berna Javier and Dr Geles Sosa, who provided diligent research assistance and good friendship. Generosity and hospitality were also extended by Romana and Eming Porcioncula, Myrna and Manny Bobadilla, Mariano Montano, Demetrio Armintia and the Colmenar family. In Manila the same can be said for Dr Helen Mendoza, Teresa Popplewell, Professor Fred Silao, Agnes Espano, Gary Auxilian and Emelyn Tapaoan. Institutional support was provided by the College of Public Administration at the University of the Philippines, Diliman, and I am grateful to Deans Proserpina Tapales and Joe Endriga, and their staff, for their kindness and support. Mayor Hermogenes Arayata of Tanza was also a gracious host.

In Vancouver, stimulation and friendship were provided by many, in particular, Terry McGee, Trevor Barnes, Robyn Dowling, David Edgington, Martin Evans, Dean Forbes, Geoff Hainsworth, Charles Greenberg, Catherine Griffiths, Nick Kontogeorgopoulos, Aprodicio Laquian, David Ley, Deirdre McKay, Scott MacLeod, Andrew Marton, Kris Olds, Jean-François Proulx, Juliet Rowson and Gisele Yasmeen.

In Singapore, I have been fortunate to land amidst a lively group of researchers and I gratefully acknowledge the support and collegial friendship of Neil Coe, Peter Dicken, Lily Kong, Kris and Andrew (again!), Elen Sia, Henry Yeung, and my colleagues in the Southeast Asian Studies Programme.

The editors of this series, John Connell, Lily Kong, and John Lea, and several commissioning and desk editors at Routledge have provided helpful comments and encouragement along the way. In the initial stages of this project, financial support was provided by a Commonwealth doctoral scholarship, the Centre for Human Settlements at UBC and the Ford Foundation through the Northwest Regional Consortium for Southeast Asian Studies. Subsequently, support was provided by a National University of Singapore Academic Research Grant (RP970013).

1

INTRODUCTION: PUTTING GLOBALIZATION IN ITS PLACE

The Philippines is no stranger to the 'global'. Centuries of contact with traders, proselytizers, colonizers and creditors have left a deep imprint on the political, cultural and economic life of the country. Such flows have also occurred in the opposite direction. An estimated six million Filipinos currently live overseas as temporary workers or migrants. Thus a sense of life lived at a larger scale than the village, town or nation is seldom absent from the everyday experiences of a large proportion of the population.

It is in this broader context for understanding the 'global' that an explicitly 'globalized' development strategy has emerged in the Philippines. Over the course of the Ramos administration (1992–1998) in particular, 'globalization' was both a key feature of the Philippine economy, *and* an important part of the administration's legitimation of its development strategies. More specifically, a broad process of deregulation, liberalization and decentralization in the 1990s has left the Philippines with an economy deliberately attuned to the needs of foreign investors in export-oriented manufacturing sectors. Implementing such political reforms has not been straightforward, but their passage has been facilitated by the constant rhetorical refrain of the necessities of globalization. As President Ramos exhorted on numerous occasions: 'There is a new reality that underscores our national life. We are part of a new global economy – in which every nation must compete, if it is to prosper. . . . [We must] imbibe and expand the culture of globalization . . . [or] be left behind in the march toward progress and prosperity for all.'[1]

While the national scale provides one context for the empirical research presented in this book, the 'landscapes' of the title are also located at other scales. In particular, attention is focused on two villages in the province of Cavite, immediately to the south of Metropolitan Manila. Cavite has received a substantial share of the foreign investment in the country during the 1990s and the result has been a booming manufacturing sector and a vibrant land market as Manila's agricultural hinterland is converted to industrial estates and residential subdivisions.

The largest single concentration of industrial investment and employment in the province (and indeed, the country) is at the sprawling 275-hectare Cavite Export Processing Zone (CEPZ) in the small coastal town of Rosario. In 1988, fewer than 1,000 people were employed in the CEPZ, and Rosario, along with the municipalities around it, were small fishing and agricultural towns. By 1998, 54,141 workers entered the CEPZ every day, coming from surrounding towns and villages. This pattern of growth has been repeated across the province at smaller scales with numerous industrial estates, many privately owned, being established. The result has been a remarkable transformation in the social and physical landscape, with major structural changes in local labour markets and the rapid conversion of agricultural land to industrial estates, residential subdivisions, golf courses and other 'urban' land uses. These towns and villages, then, represent the landscapes of globalization in their most literal sense. Two villages in Cavite are used in this study to explore these processes in detail. In one, agricultural activities – largely wet rice cultivation – continue but under considerable financial and social strain, as the labour market, especially among the younger generation, reorients itself towards non-agricultural forms of work. In the second village, agriculture has been virtually eradicated as residential developments and land speculation drive tenant farmers out of business.

These material changes, then, form the social and physical landscapes of globalization that this study sets out to describe. There is however, another looser sense in which 'landscapes' of globalization inform the argument of this book. Globalization exists not only as a set of material processes of economic and social change, but also as a set of ideas and discourses that foster a specific representation of the global economy. Here, the issue of scale is central. To represent economic and social change as a consequence of 'global forces' and to justify development strategies with reference to the requirements of such forces (as Fidel Ramos does in the quote above), is frequently to defer the understanding, explanation and meaning of socio-economic change to a scale that cannot be easily grasped or held accountable. The global scale is all too often represented as a disembodied set of forces that create an inevitable and inexorable context for contemporary political reality. What we are witnessing therefore is a production of scale in which the global is constructed as the fundamental level at which social and economic change should be understood.

Globalization, then, is more than simply a set of material processes – it is also a representational strategy in which the social world is rendered comprehensible. In essence, it creates a particular spatial imaginary that invokes a global space of flows in which specific places are inserted. This, then, is the more abstract 'landscape of globalization' – the imaginary landscape of the global economy in which people and places construct their 'situation' in global space. Such a construction is far more than just imaginary, however, as it informs and legitimizes developmental experiences such as those that will be described in the villages of Cavite and elsewhere in the Philippines. The imperatives of global competition and a globalizing context for development have been a central part of the Ramos administration's strategy for national development. They have been used to justify deregulation, decentralization, and liberalization, and to allow foreign direct investment in infrastructure, export manufacturing and service industries. It is these policies that have provided the national and provincial context for socioeconomic transformation in Cavite, and thus it is important to explore both their discursive roots and material consequences.

This introductory chapter will serve to spell out the questions that inform the study ahead and the arguments that will be developed. To begin, I will outline the nature of material processes of globalization and suggest reasons why they have drawn such attention in the last decade in particular. This forms part of a discussion of what might be termed the 'logics' of globalization – the theoretical frameworks that indicate the various dimensions of the globalization process. These theoretical constructions provide some explanations of the material processes that are usually included under the rubric of globalization. At the same time, they illustrate the way in which the global has been discursively constructed in the academic literature as having a logic of its own. In the second section, I introduce some recent ideas on the production of scale to suggest that these 'logics' of globalization can be viewed as social constructions of scale, rather than providing objective divisions of space or *a priori* levels of analysis. This forms the basis for arguing that 'globalization' must be seen as a process that operates at multiple scales and is therefore embedded in place-specific social relations. The third section of the chapter argues that the social construction of scale must also be seen as political because it has been used to legitimize a neoliberal approach to economic development. It is in this sense that I will later refer to the political discourse of globalization that is deployed at national and other scales.

The logics of globalization

The process of globalization is usually taken to mean the increasing porosity of physical and social barriers to world-wide flows of capital, goods, people, ideas, imagery and institutions.[2] It is found in the flows of capital between the world's financial centres ('hot money'), and in flows of foreign direct investment.[3] It is found in the transfer of goods and services across global space – from Coca-Cola to coconuts and from microchips to missiles. It is found in the passage of people between places – tourists, refugees, migrants, and contract workers. It is found in the ideas and information that pass freely, or sometimes not so freely, across space – everything from environmentalism and human rights to neoliberalism and organized religion. It is found in the imagery to which many are exposed – Hollywood, high fashion, and hard rock. Finally, it is found in the political institutions that bring together sovereign territories into broader frameworks: the World Trade Organization, the United Nations, and the International Monetary Fund. Thus, in his trilogy describing a new 'informational' global economy, Manuel Castells argues that

> our society is constructed around flows: flows of capital. Flows of information, flows of technology, flows of organizational interaction, flows of images, sounds, and symbols. Flows are not just one element of the social organization: they are the expression of the processes *dominating* our economic, political and symbolic life.
> (Castells, 1996: 411–12; emphasis in original)

Globalization is, however, about more than flows. As several writers have noted, the contemporary intensity of flows in the global economy can be shown to have equally impressive historical precedents (Hirst and Thompson, 1996). Rather, globalization is also a shorthand means to convey a sense of the interconnection and integration of activities across the planet (Dicken, 1998). Economic activities and spaces no longer relate to one another simply through transactional linkages, but rather they are increasingly entrained in integrating networks of coordination and control. It is therefore functional integration rather than connectedness *per se* that makes the present distinct.

What drives this process of flow and integration? We might identify four 'logics' that have been used to understand the processes of globalization. The first of these is an economic imperative rooted in an understanding of the logics of the capitalist mode of production. One of the most authoritative and influential voices on this matter is David Harvey, through his project to develop a spatialized marxism, or historical geographical materialism. Harvey's argument is based on an understanding of three fundamental characteristics of the capitalist mode of production: (1) it is growth-oriented in terms of output and value; (2) growth in real values rests upon the exploitation of living labour in production; and, (3) it is technologically and organizationally dynamic in the search for profit (Harvey, 1989: 180). Arguing that these imperatives are ultimately contradictory, it was Marx's insight that capitalism is prone to periodic crises of overaccumulation. Harvey shows that of the systemic choices available to manage these crises, the temporal and/or spatial displacement of overaccumulation is perhaps the most promising (Harvey, 1982). This entails the creation of new spaces for capitalist production across the globe and it is these structural processes which underpin the 'time–space compression' characterizing the condition of postmodernity from the early 1970s to the present (Harvey, 1989). As the Fordist regime of national economic regulation broke down in the industrialized core, so strategies of time–space displacement drew the global periphery ever more closely into the world economy. Harvey concludes that if there is anything distinctive about the current nature of capitalism it is the role of financial flows and credit, and, if there is to be any stability in the current regime of accumulation, it will derive from further rounds of temporal and spatial 'fixes' (Harvey, 1989: 196).

A second set of logics driving globalization is *technological* change. To separate a technological 'logic' from economic imperatives is in some ways a misrepresentation, but the distinction is worth making because a number of writers strongly emphasize the importance of technological change in creating and facilitating both economic and cultural flows at a global scale. The most notable proponent of this view is Manuel Castells (1989, 1996, 1997, 1998). According to Castells, a new technological paradigm emerged in the two decades between the late 1960s and the late 1980s. The scientific and technical core of this paradigm lies in microelectronics, but it also includes the application of these technologies to telecommunications and biotechnology (Castells, 1989: 12; 1996). Such technologies have created a period of time–space compression in which distant places can be materially and virtually connected far more easily and economically. This has allowed an acceleration and increased volume in flows which already existed, the emergence of new flows, and the functional integration of activities across global space (Castells, 1996).

In his account of the reformulation of capitalism, Castells follows the same crisis-resolution argument as Harvey. The point at which Castells diverges from Harvey, however, is in seeing technological change and the informational mode of development not as responses to crisis but as a condition for capitalist restructuring:

> Advances in telecommunications, flexible manufacturing that allows simultaneously for standardization and customization, and new transportation technologies emerging from the use of computers and new materials, have created the material infrastructure for the world economy, as the construction of the railway system provided the basis for the formation of national markets in the nineteenth century.
> (1989: 30)

For Castells, then, we have entered a new informational age, facilitated by late-twentieth-century technology:

> The informational economy is global. A global economy is historically new, distinct from a world economy. A world economy, that is an economy in which capital accumulation proceeds throughout the world, has existed in the West at least since the sixteenth century.... A global economy is something different: it is an economy with the capacity to work as a unit in real time on a planetary scale. While the capitalist mode of production is characterized by its relentless expansion, always trying to overcome limits of time and space, it is only in the late twentieth century that the world economy was able to become truly global on the basis of the new infrastructure provided by information and communications technologies.
> (Castells, 1996: 92–3)

The corollary for Castells (1996: chapter 6) is the emergence of a *space of flows* dominating the historically constructed *space of places*.

A third set of logics, revolving around *political* institutions, might be represented through the work of Anthony Giddens. For Giddens (1990, 1991), the process of globalization is linked to a longer process of modernization, where modernity has four central characteristics: capitalism; industrialism; surveillance; and state control over the means of violence. The process of modernization, starting in the eighteenth century, involves three changes in the way in which social life is experienced (and, following Giddens' ideas of structuration, the experience of individual agents is reflexively linked with social change). The first change is the distanciation, meaning separation, of time from space. From diurnal and seasonal rhythms, time became standardized and universalized in the eighteenth century through the mechanical clock. Similarly, space was no longer an experientially perceived space, but could also be universalized in maps. These developments allowed the universalization of human activities across spatial and temporal distances. A second change associated with modernity is the 'disembedding' of social relations from local contexts, through 'symbolic tokens' (particularly money) and 'expert systems' (for example, technical knowledge). Thirdly, an awareness of the role played by information and knowledge, and the risk involved in trusting expert systems, make modern individuals, according to Giddens, particularly reflexive in a way that human subjects have not been in the past.

Giddens' three characteristics of modernization – time–space distanciation, disembedding, and reflexivity – promote universalizing tendencies that render social relationships across space more inclusive, thereby allowing globalization of social interaction. This is the basis upon which Giddens can proceed to explain the current trends of globalization in terms of his dimensions of modernity (capitalism, surveillance, military order, and industrialism). Capitalism has constructed a world system, as discussed by Wallerstein (1974), incorporating the globe into a single market for commodities, labour and capital. The globalization of production involves the spread of industrialism and diffusion of technology around the globe and the incorporation of local production systems and labour markets into the international division of labour. Surveillance has also been globalized by international organizations of nation-states. Finally, military globalization is incorporated into a (precarious) 'global' alliance system (and, after Giddens wrote, into a so-called 'New World Order' focused on the UN Security Council) (Waters, 1995).

In short, Giddens provides a multi-dimensional explanatory framework for understanding globalization processes, but the main thrust of his argument seeks to balance the economistic theories of writers such as Wallerstein with a consideration of the role of the nation-state and military/political order in generating globalized structures and subjectivities: 'it is surely plain to all, save those under the sway of historical materialism, that the material involvements of nation-states are not governed purely by economic considerations, real or perceived' (1990: 72).

The fourth logic of globalization is *cultural*. While many would be resistant to any suggestion of a cultural *logic* at the global scale, with the implication of some systemic inevitability, there is, nevertheless, a 'logic of flows' identifiable in the work of cultural theorists and global sociologists (Featherstone and Lash, 1995: 23). We can identify two approaches to the global logic of culture, one emphasizing the role of the *imaginary* and its construction at a planetary scale, the other concentrating on the role of human *agency* in the global spread of common cultural meanings and significations.

Sociologist Roland Robertson exemplifies the view that globalization is a product of the social imaginary, such that 'there is a general autonomy and "logic" to the globalization process – which operates in *relative* independence of strictly societal and other more conventionally studied sociocultural processes' (Robertson, 1990: 28). Robertson provides a two-part definition of globalization as: 'the compression of the world and the intensification of consciousness of the world as a whole' (1992: 8). While the first part of this definition shares common ground with the likes of Harvey and world systems theorists, Robertson chooses to distinguish his position by emphasizing the rise of global consciousness, through a variety of events and institutions: 'conceptions of the world-system, including symbolic responses to and interpretations of globalization, are themselves important factors in determining the trajectories of that very process' (Robertson, 1992: 61).

Robertson's ideas finds considerable resonance in Arjun Appadurai's (1990) influential work on global flows. Appadurai too focuses on the role of the human agent and the social imaginary in constituting globalization by identifying five sets of flows, or 'landscapes' (based on people, finance, technology, media and ideas). He emphasizes that

> the individual actor is the last locus of this perspectival set of landscapes, for these landscapes are eventually navigated by agents who both experience and constitute larger formations, in part by their own sense of what these landscapes offer. These landscapes thus, are the building blocks of what . . . I would like to call 'imagined worlds', that is, the multiple worlds which are constituted by the historically situated imaginations of persons and groups spread around the globe.
>
> (Appadurai, 1990: 296–7)

A second, and more systemic, way of viewing globalization in a cultural frame is to focus on the actors and agents involved in the production and diffusion of 'global cultures'. Thus for Friedman, 'in global terms, the culturalization of the world is about how a certain group of professionals located at central positions identify the larger world and order it according to a central scheme of things' (Friedman, 1995: 82). In this way, what Friedman describes as Robertson's overly 'mental and semantic' approach is recast by situating actors in their political and economic contexts (Friedman, 1995: 72).

Several writers, particularly those concerned with 'culture industries', have pursued Friedman's point about a system of agency – or a network of actors. Some focus just on a small elite cadre of 'cosmopolitans' (Hannerz, 1990) or 'third cultures' (Featherstone, 1990: 7), such as diplomats, financiers, academics, aid officials, and media representatives, who act as cultural intermediaries. These groups, according to Featherstone, are establishing 'sets of practices, bodies of knowledge, conventions and lifestyles' increasingly independent of national origins (Featherstone, 1995: 114). Others, however, engage more broadly with the role of individuals involved in generating a globalized set of 'signs'. These actors, operating in sectors such as advertising, music, tourism, publishing, design and the media represent the 'tendential' beginnings of a process in which information, communication and signs form the building blocks of what Lash and Urry (1994) call 'economies of signs and space'. In this world a 'logic of flows' replaces a 'logic of organizations' (Featherstone and Lash, 1995: 23).

These four logics, then – economic, technological, political and cultural – have been identified as constituting the process of globalization. The reason for highlighting these accounts is, however, not simply to provide a literature survey of approaches to globalization. The logics identified, often critically, by these authors have been adopted in far less critical terms by popular writers who represent globalization as an irresistible force to which people and places must submit themselves. Kenichi Ohmae, for example, talks of 'putting global logic first', and provides a stark choice: 'if a country genuinely opens itself up to the global system, prosperity will follow. If it does not, or if it does so only halfheartedly, relying instead on the heavy, guiding hand of central government, its progress will falter' (Ohmae, 1995a: 123). Implicit in Ohmae's argument are two fundamental assertions: first, that the globalization of economic activities is inevitable and inexorable; and secondly that only complete compliance with the requirements of global capital will bring prosperity to any particular place. This is the discursive context in which politicians and policy-makers concerned with economic development currently operate. It is, furthermore, a view that is represented as objective and apolitical. John Naisbitt, another guru of globalized business strategy, argues that with free trade and economic liberalization, 'ideology is giving way to economic and political reality' (1995: 121).

The popular representation of globalization and especially economic globalization, then, is as the determining context for action, and one to which individual governments (and other institutions) must open themselves and adjust. The assumption is that allowing such flows to pass free and unhindered between places in the world will ultimately be unequivocally beneficial for all concerned. The relationship of places, then, to a 'space of flows' is one of both 'dependence and vulnerability' (Castells, 1996: 384):

> In this network, no place exists by itself, since the positions are defined by flows. Thus, the network of communication is the fundamental spatial configuration: places do not disappear, but their logic and their meaning become absorbed in the network. The technological infrastructure that builds up the network defines the new space. . . . The space of flows is not placeless, although its structural logic is.
>
> (Castells, 1996: 412, 413)

Globalization and the relativization of scale

The logics of globalization, then, provide a foundation for believing that the world is moving unavoidably towards ever greater interconnection. They suggest that in various spheres – economic, technological, political and cultural – all peoples and places are being inexorably manoeuvred into a globalized frame. This has important implications for how we understand the relationship between particular places and the global economy. In Castells' terms, a space of places is being subverted by a space of flows. More prosaically, the global scale in general becomes the scale at which everyday life is being rendered comprehensible. Industrial policies, developments strategies, welfare frameworks, labour laws, environmental regulations and so on must, in a globalized imaginary, be tailored to the perceived or constructed needs of exogenous forces of globalization. We may not all be directly 'wired' in to the global informational economy – and certainly large areas and vast numbers of people are not – but according to Castells, 'all economic and social processes do relate to the structurally dominant logic of such an economy' (1996: 103).

What is occurring here, I would argue, is a complex conflation of material and discursive processes. To consider firstly the material dimensions of globalization, it is true that important processes of change are occurring in all of the four spheres described above. Qualitative changes have taken place over the last three decades in the global economy: production systems have integrated distant sites of activity; technology has profoundly transformed all aspects of life; political integration has not brought about global government, but institutions of global governance have certainly emerged with considerable power; and cultural interaction and hybridization have become increasingly possible. Social relations across space are therefore being integrated in more intensive and extensive ways.

The corollary of these processes is that space is being experienced in distinctively different ways than in the past. Several authors have attempted to capture this idea through the concept of the 'relativization of scale.' Essentially, this means that geographical scales, such as the nation-state, the city, the region, etc., cannot be considered fixed containers in which social processes occur. Instead, according to Jessop,

> we are witnessing a proliferation of spatial scales . . . their relative dissociation in complex tangled hierarchies (rather than a simple nesting of scales), and an increasingly convoluted mix of interscalar strategies as various economic and political forces seek the most favourable conditions for their insertion into the changing international order.
> (Jessop, 1999: 24)

The result is a confusing world in which many structural certainties have evaporated and spatial categories have become far less meaningful. It no longer makes so much sense, for example, to devise economic policy at a national level. Instead, development strategies are now being played out at the scale of the city, the nation, the region and the global all at the same time. Castells neatly captures this chaotic collapse of inherited economic geographies:

> The global economy emerging from informational-based production and competition is characterized by its *interdependence*, its *asymmetry*, its *regionalization*, the *increasing diversification within each region*, its *selective inclusiveness*, its *exclusionary segmentation*, and, as a result of all these features, an extraordinarily *variable geometry* that tends to dissolve historical, economic geographies.
> (Castells, 1996: 106; emphasis in original)

9

This variable geometry, as Castells calls it, forms a new spatial division of labour in which old hierarchies, territories and spaces are being reworked into a new global economic architecture. In the terminology of Thrift and Olds (1996), the topological presupposition of the 'bounded region' is being superseded by that of the 'network'.

Until relatively recently, space, as the material venue for 'time-sharing' social practices, could generally be assumed to represent the contiguity of such practices. Now, in a networked space of flows, simultaneous practices are linked at a distance and geographical proximity no longer has the same meaning (Castells, 1996: 411). Spatial scales – nation, region, city, etc. – no longer contain social practices, but instead such processes may now occur simultaneously at multiple scales. Any suggestion that this is a universal characteristic of social life across the world is clearly ludicrous, but what is being identified here is a prevalent tendency, rather than a completed end-state.

The response to this confusing, chaotic, relativization of scale brings us to the discursive dimensions of the question. Rather than recognizing the multiplicity of scales at which social processes are simultaneously constituted, the reaction has too often been instead to *privilege* the global scale in a hierarchy that implies its priority in determining social outcomes at other scales. This, I would argue, is precisely what has happened in the popularization of globalization as an idea. We have witnessed a popular representation of scale in which the global is seen as paramount and is imbued with determinative influence.

The key process here is the *production of scale* – the creation of a level of resolution at which phenomena are deemed understandable (Kelly, 1997, 1999b).[4] This notion can be illustrated with two examples. Firstly, the 'miracle' of rapid industrial growth in East Asia might be explained in terms of a 'Confucian' culture of individual entrepreneurialism and hard work (the scale of the body), the proactive role of the state in directing growth (the national scale), or, the operation of free market forces and open trading relationships (the global scale). Each scale provides a quite different perspective on the issue and a different set of political judgements. Secondly, while the proximal cause of the Asian financial crisis – the 'end of the miracle' – might have been the over-exposure of national economies to short-term, unhedged, foreign currency loans, the root causes of this state of affairs can be explained at a variety of scales: the body (the individual actors, such as George Soros, blamed for orchestrating capital exodus); local (firm strategy); national (regulatory structures; cronyism); regional (rapidly developing economies; bull markets; contagion) and global (aggressive international banking practices in the context of deregulation, competition and overaccumulation; the 'moral hazard' of IMF bail-outs) (Kelly and Olds, 1999).

But if, as I have argued, we see scale as relativized under the contemporary processes of the global economy, then there are significant implications for this way of thinking about globalization. Rather than coming to terms with the ways in which globalization as a determinative process has 'local', 'regional' and 'national' impacts, it should instead be recognized that social processes happen at none of these scales exclusively but at all scales simultaneously. Thus, no scale can be claimed as the privileged level for explanation (Swyngedouw, 1997). Expressions such as the local–global dialectic or 'glocalization' are therefore helpful only in the extent to which they assist in the collapsing of such a dualism. The local and global, or place and space, should thus be seen as dialectically related (Merrifield, 1993; Swyngedouw, 1997).

To highlight globalization as a discursive construction in this way is not then to deny the material processes and changes that are occurring. Instead, it is to suggest that they have been misunderstood as hierarchically ordered with 'global processes' occupying the highest level. But the privileging of the global scale is not accidental. Producing scales is not a politically neutral act, even though the pervasiveness of this understanding across the political spectrum makes it appear so. Explaining a phenomenon at the global scale is a political judgement not a technical one: 'scalar narratives provide the metaphors for the construction of "explanatory" discourses. . . . Scale is, consequently, not socially or politically neutral, but embodies and expresses power relationships' (Swyngedouw, 1997: 140). While Castells (1996) recognizes that the globalizing network economy that he describes involves great inequalities between those with the power to control, and intercede between, networks, what he fails to acknowledge is the discursive power of representing economic life in this way in the first place.

What, then, are the politics of globalization and 'global logics'? It would be wrong to suggest that the construction of a global scale is a conspiracy among those with a particular ideological leaning or a powerful few who stand to gain. Instead, globalization is a construction in which many are complicit and the diversity of the authors cited earlier in this chapter is a testament to that fact. Nevertheless, as Agnew and Corbridge argue, the idea of globalization creates a Lefebvrian 'representation of space' in which it is primarily neoliberal economic, social political practices that are legitimized (1995: 204; see also Kelly, 1999b).

Globalization serves as a particular discourse of development that is used to justify neoliberal economic policies in which the state is viewed as a hindrance to economic development. Globalization gives authority to neoliberal arguments for market access and free trade – what has been called the 'counter-revolution' in development theory. It creates the context in which openness to the global market is seen as the inevitable and common-sensical route to prosperity and progress – what Broad and Cavanagh call the 'Washington consensus' (1993: 156). Broadly speaking, this neoliberal orthodoxy holds that

> markets offer a guarantee against the corruptions of government (and Leviathan), and they embody the most reasonable way of dealing with, and making sense of, a world based around fluidity, flows, change and movement. States, in this discourse, are about stasis, sedimentation and distortions; markets offer an antidote to such self-willed sclerosis and entrenched hegemonies.
>
> (Agnew and Corbridge, 1995: 200)

The 'counter-revolution' in development theory has brought the 'principle' of the free market to the centre of development economics (Toye, 1993). The work of neoliberal theorists has created, or perhaps merged with, a consensus in policy circles around the 'fundamental' objectives of fiscal discipline, tax reform, financial liberalization, privatization, deregulation, and support for property rights. In terms of domestic economic management, this translates into policies to reduce price controls and subsidies, eliminate state marketing boards, and limit wage indexation. The key component of the consensus, however, relates to trade liberalization. This involves policies such as exchange rate flexibility, eliminating trade licensing systems, providing export incentives, liberalizing conditions for foreign investment, reducing tariffs and eliminating quota restrictions (Biersteker, 1995: 177–8).

It is not, however, simply disembodied 'global capital' that promulgates such a view. As Biersteker points out, 'local' vested interests play a fundamental role in promoting neoliberal policies. In other words, development strategies cannot be divorced from their political economic context, and all politics, as the saying goes, is local. It is, then, at subglobal scales that the discourse of globalization and the practices associated with it are played out, and this forms a fundamental point developed in this book. Far from Castells' argument that places now exist only in relation to global flows, my argument is that globalization, both as material processes and discursive construction is subjected to significant local mediation. Thus processes ascribed to 'global forces' should actually be seen as constituted far more prosaically.

The arguments

The central question that guides this inquiry is how social processes embedded in places at multiple scales, *mediate* and *construct* a particular experience of globalization.[5] In order to address this question, several arguments will be developed throughout this book. Firstly, the 'places' in which globalization is embedded are at many different scales – national, regional, provincial, municipal, village, household and individual. These scales cannot be seen as distinct and hierarchical, with social processes at one scale determinant over another. Instead, they are simultaneously interlinked and overlapping. The global exists only in multiple 'locals' and the local and the global are not 'natural' and distinct scales of analysis but are, instead, mutually constitutive of each other. Globalization, then, is not some *deus ex machina* to which political choices have to be deferred; instead, it embodies a set of processes and human actors with motives and agendas that are considerably more parochial.

In this way, I wish to distance myself from such writers as Castells who suggest that a space of flows exists independent of, and dominant over, a space of places. Castells summarizes his position as follows:

> Thus, people do still live in places. But because function and power in our societies are organized in the space of flows, the structural domination of its logic essentially alters the meaning and dynamic of places. Experience, by being related to places, becomes abstracted from power, and meaning

is increasingly separated from knowledge.... The dominant tendency is toward a horizon of networked, ahistorical space of flows, aiming at imposing its logic over scattered, segmented places...

(1996: 428)

But here Castells is reproducing precisely the production of scale described earlier – the dominance of the global over the local; space over place. My argument here is that while important changes are occurring in material processes in the contemporary global economy – and this includes many of the dimensions of the network society that Castells identifies – they are constituted and experienced at multiple, relativized, scales.

The second key argument is that while globalization, and its embeddedness in places, operates as a material process based on time–space compression, it is also a social construction and political discourse. In other words, globalization is *both* a *real* phenomenon that is experienced by people in the Philippines, but it is also an *idea* that carries with it powerful implications for the geography, experience and social justice of development. As a political discourse, globalization serves to legitimize certain practices and construct a particular relationship between 'global spaces' and 'local places'. Thus the material processes of globalization create a discursive frame around them, but equally, material processes are called into being by the discourse that constructs them. As Piven notes, globalization 'has become a political force, helping to create the reality that it purportedly merely describes' (1995: 108). Thus appeals to 'embrace' globalization as inevitable and unavoidable are not the realism that their rhetoric implies, but rather political judgements that can be deconstructed.

The logics of globalization described earlier highlight processes that, for practical purposes, do set a context that cannot be chosen or readily altered. The constitute, as the politicians and gurus of globalization suggest, an unavoidable backdrop for policy-making and action. But, to paraphrase Marx, what is lacking in many contemporary representations of globalization are the people, or even institutions, who are creating history, but not in conditions of their own choosing. This, I believe is the point at which an intervention is needed. The leap in logic from the *context* provided by globalization to the *necessities* that it implies must be questioned, because it is here that the construction of the global scale becomes politicized.

Governments – and in this study that means the various levels of administration in the Philippines – have readily subscribed to the supposed imperatives of globalization and elided them with a neoliberal agenda of free trade and economic liberalization. Globalization is not synonymous with neoliberalism, but it has been constructed as such – it has become part of a discursive legitimation for thinking about the relationship between particular places and global space. In this sense, globalization is a political discourse serving to naturalize development policies that should be contested and questioned.

The basis of this discourse is a production of scale in such a way as to make the relationship between places and global space understandable. It encourages us to think about our 'places in the world' in a particular manner. Specifically, the appeal to the notion of globalization enframes places as 'nodes' in a 'network' of global flows in which states, places and actors are denied any proactive role. Social processes that operate at the *local* scale (whether this be national, provincial, village, etc.) are represented as subordinate to the *global* scale. In other words, scale, particularly the global scale, is being socially constructed in the deferral of political options to the necessities of globalization.

Here again, my argument diverges from Castells' work on globalization. Castells sees power in a globalizing network society as the ability to construct links between networks, the 'switchers' as he calls such intermediaries – 'the switchers are the power holders' (1996: 471). I would argue, however, that power is not defined simply by the ability to harness global networks in this way. Firstly, power is also embedded in the ability to intermediate in networks at smaller scales as suggested above. For every elite global 'switcher', there are scores of other power holders who at smaller scales intermediate experiences of globalization. This study will highlight how such intermediaries have operated in specific contexts in the Philippines. As Vicente Rafael argues, power is frequently a product of the ability to broker relationships between the 'inside' and the 'outside' – to 'lay claim over the site of circulation' (1995: 5). Secondly, as noted above, appeals to the inevitability and dominance of 'global forces' are political strategies that serve to legitimize a structure of power relations at smaller scales. Thus power holders in particular political economies are able not just to mediate globalization but also to construct or represent it in such a way as to entrench their own positions.

Structure of the book

Two key points, then, guide this book: that globalization is a material process embedded in the social relations of particular places across multiple scales; and, that globalization can be seen as a social construction that has been deployed for political purposes. If globalization is interpreted in this way, we can start to destabilize the sense of *inevitability* that envelopes it, and begin to recognize that it is a construction of scale that has political implications which need to be scrutinized.

Chapter 2 initiates these arguments by tracing the historical roots of Philippine relations with 'global space' from pre-colonial times to the present. The account shows that this relationship has been contingent and politically contested over time and has owed as much to domestic power relations as to global forces. In making this point, the chapter also lays out some of the historical roots of the social, economic and cultural structures that form the place-based context in which contemporary globalization is embedded.

Chapter 3 describes the imprint of global capital in the Philippine space economy. Recent experience shows that a strategy of globalized development creates a spatial pattern of growth that reinforces existing inequalities and focuses on core regions most closely integrated with global transactional flows. The result is a pattern of mega-urban development that has characterized the Philippines and other newly industrializing countries in East and Southeast Asia. The reasons for this geography are, however, universal only in a superficial way. Specific political circumstances at the national and subnational scales have played a key role in shaping the Philippine space economy.

Chapters 4 and 5 bring the discussion to the scale of two 'globalizing villages' in the heart of the transforming landscape around Manila. Here, the significance of locally embedded social processes is highlighted in exploring experiences of globalized development. This point is made with respect to the two key dimensions of globalization in Cavite: the transformation of local labour markets, and the conversion of agricultural land to residential and industrial uses. In the village of Bunga, a local labour market transformed by new industrial employment is considered in the context of both local political practices regulating the industrial labour process, and the agricultural production process where the implications of recent changes are most acutely felt. In the second village, Mulawin, land is the factor of production under analysis and once again experiences of globalization are shown to be embedded in locally constituted relations based on political power, social norms and environmental processes. In both cases, the economic and cultural dimensions of globalization are shown to be integrated in a process of hybridization between existing meanings of gender, youth and work, and the new horizons offered by globalization.

Having established the ways in which globalization is embedded and implicated in local power relations, we then move on in chapter 6 to consider how it has been resisted and rethought in the Philippines. At the discursive level, globalization has been re-imagined through the writings and activities of various intellectuals, politicians and organizations that have questioned the dominant construction of the Philippines' place in global economic space. In a more practical sense, the local consequences of globalization have been resisted through both individual and collective acts of subversion. Such resistance has, however, tended to be successful only when conscious of both the multiple scales at which globalized development is constituted, and the social power relations in which it is embedded. The implications for 'local' resistance to the 'global' are discussed and I argue that given the evidence of locally constituted experiences of globalization, to talk of 'local' resistance is meaningless, since the globalization process itself is already local.

2

THE PHILIPPINES AND THE GLOBAL ECONOMY

Constructing a place in the world

This chapter explores the foundations of a globalist discourse in the Philippines – how a distinct sense of the country's place in the world has been shaped by a history of trade, colonialism, and proselytization. It then considers how this construction of a Philippine place in the world has translated into policies and patterns of development.

Several key points emerge from this historical contextualization of the present. Firstly, global processes are mediated and constructed at multiple scales and through various agents. The historical experience of globalization in the Philippines cannot be understood in terms of exogenous forces bearing down upon a passive and pliable site. Instead, traders, colonizers, missionaries and multilateral institutions have always had to work through institutions and actors at other scales – national, provincial, municipal, village and so on. Furthermore, these actors and institutions have always been proactive in their response to supranational influences and processes. At various times, the support and legitimacy provided by religious and colonial hierarchies has proved useful to certain parties as national and local level power relations have been played out. As we will see in this chapter, the hierarchical structure of race and power under Spanish colonialism served the interests of the Filipino elite, while the institutions of local democracy established by the American regime further embedded and legitimized this power structure. Later, the influx of capital from loans orchestrated by the IMF and World Bank proved to be a key support for the authoritarian regime of Ferdinand Marcos.

A second point that emerges in this chapter is that while the Philippines has long been integrated with the world economy, the nature of this relationship has shifted significantly in the last few decades. Changes have occurred within the country, but just as importantly, significant transformations have taken place in the processes and institutions that comprise the world economy. These changes are multifaceted and not easily captured in any universal theoretical framework, but they represent a marked intensification in the connectivity of global economic space and the extension of that space to include previously marginal areas. Thus, the global economy is becoming increasingly integrated both quantitatively in terms of flows of capital, commodities, people, information, etc., and qualitatively in terms of the dimensions in which this integration is being experienced. Globalization might be a rather weak 'catch-all' to describe these changes, but there is little doubt that it connotes some important processes in the contemporary world economy.

A third issue that emerges serves to deflate the political power of globalization as an idea. Here I refer to the discursive nature of globalization as a metaphorical construct to characterize the global economy. As the later sections of this chapter demonstrate, globalization has become a powerful tool for political and business interests to justify certain social and economic policies.

The construction of globalization in the Philippines is a product of historical experience rather than an instant creation and so a fourth theme that emerges in this chapter is the historical process through which powerful actors have produced the country's 'place in the world'. The construction of globalization and its implications for the Philippines have been produced within the country's power structures rather than imposed from outside or derived purely from some politically neutral assessment of the external economic environment. Furthermore, the legitimacy of this construction of the Philippines' place in the world is rooted in the historical experience of interaction with global processes. In other words, this chapter makes the rather bold assertion that the country's historical context has left it peculiarly susceptible to arguments that prioritize the global scale and that this is inscribed upon the landscape in the types of development experienced. The boldness of this suggestion is that it creates an explicit link between the cultural legacies of religious conversion, colonialism and intense trading links, and the contemporary context for economic policy.

In making these points, this chapter is structured chronologically, starting with the pre-colonial society that the Spanish encountered, followed by the experience of Hispanic colonialism (1521–1896), then the American period (1898–1946) and finally the post-war (and post-independence) Philippine state. Beginning with a description of the pre-colonial Philippine archipelago is important because it draws attention to distinctive elements of Filipino cultural development sometimes forgotten in accounts of colonialism. This is not to say that there is a foundational Filipino identity to be unearthed, but it does highlight the fact that subsequent outside influences encountered a culturally complex society that already had substantial contact with the outside world. The processes of colonialism were, therefore, closer to hybridization than assimilation.

The nature of this hybridization under Spanish rule will be discussed in several areas: trade and production; social structure and government, and religion. In the case of trade and production, we see how economic interest groups, such as Spanish merchants and colonists, and British and American trading houses, vied to define the nature of the colony's articulation with the world economy in order to advance their own interests. In the realms of government, social relations and religious hierarchies, I will suggest that the significant effect of colonialism was less a reworking of social and cultural frameworks – most were kept in place in a modified form – but more their formalization and expansion so that hierarchies concluded not with local chiefs or shamans but with authorities far removed. It is in this process – the colonization not of territory or resources, but of consciousness – that the roots of a globalization discourse can be located.

The American experience in the Philippines also worked on both economic and cultural levels. Through an ostensibly more benign, but perhaps therefore more insidious, brand of colonialism, the USA sought to advance both its own interests and those of a particular class of Filipinos. As an exogenous influence, US colonialism represented an input to a pre-existing structure of local power relations that left hierarchies and hegemonies even more entrenched. The post-independence era saw a similar pattern persist, with outside support being used for domestic political purposes. At the same time, domestic political constituencies have multiplied to create a variety of views with regard to Philippine relations with the outside world. It is those constituencies that have jockeyed for influence in recent years and have arrived at the political economy of globalization that now holds sway.

The pre-colonial islands

When, in 1521, Ferdinand Magellan arrived and claimed for the Spanish crown what would become the Philippines, he found an archipelago whose political structure bore no relation to the contemporary Philippine state and indeed where the nation-state was an unfamiliar concept. No unifying pre-colonial empire existed as it had elsewhere in the Malay world, for example under Majapahit (centred on Java) and Srivijaya (on Sumatra). Instead, there existed a system of local sultanates and fiefdoms controlling limited hinterlands from coastal and riverine settlements (Constantino, 1975). Different cultural, linguistic and social systems existed in the archipelago's various regions – differences that still carry some weight today in the construction of regional identities, for example as Ilocanos, Tagalogs, Bicolanos or Visayans (and extending to subgroups within such regions).

Rice was then, as now, the staple crop, but swidden agriculture also yielded root crops. In a few areas complex, labour-intensive irrigated field systems had developed for rice cultivation, although seasonal flooding also provided natural irrigation. Prior to Spanish missionization neither ploughs nor draft animals were used, and land was plentiful with large areas still uncultivated. Fish formed the other main staple, with coastal waters and rivers exploited for this purpose. Craft production was well developed by the sixteenth century and while local goods were for subsistence purposes, surpluses would be traded with other communities.

By the time the Spanish arrived, Manila was already the archipelago's major entrepôt port, acting as a centre for trade between other islands and the rest of the region. Merchants from Borneo, China, Japan, Siam, Cambodia, India and the Islamic and Malay worlds frequented Manila and had done so for several hundred years (Reid, 1993: 60). Some also settled in Manila – by 1571, when the Spanish landed in Manila, there were 150 Chinese in residence among a population of about 2,000 (Caoili, 1988; Yoshihara, 1985).

Exports from Manila were mostly primary products such as wax, honey, leather, deerskins, raw cotton, palm wine, and gold, but given the port's significance as a trading centre, foreign goods such as Chinese silks and porcelain were also traded there. Imports to Manila were predominantly manufactured items such as textiles, crockery, kettles and swords, and commodities such as copper, pepper and precious stones. Such goods were exchanged with other settlements in the Southern areas of Luzon and to a limited extent throughout the archipelago via inter-island or upland–lowland trade routes (Caoili, 1988).

The focus of trade on Manila did not, however, imply any form of political control emanating from the growing core region. Instead, dispersed *barangays* (villages) of 100–500 people were mostly engaged in subsistence cultivation and formed self-contained fiefdoms. The word *barangay* itself indicates something of the nature of these communities. Meaning 'boat' in Tagalog, the word refers to the initial settlement of the islands by individual boatloads of migrants and implies the close ties that bound members of the same community, through kinship, allegiance or alliance (Reid, 1988).

The 'captain' of a village was a *datu*, and together they formed an aristocratic *maginoo* class in the Tagalog region around Manila. Some villages were grouped together as a *bayan* or town, with one *datu* taking precedence over others by virtue of superior economic or military power. The *datu* would act as the military, political and legal chief and could command services, agricultural produce and respect from his people. Two of the most powerful of these rulers were to be found in the settlements of Tondo and Maynila, at the mouth of the Pasig River where Manila is now situated.

The *datu* class formed one part of a three-tiered social system that was essentially feudal, but with important differences from the European equivalent at the time. Beneath the *datu*, there was a class of 'freemen', called *timawa* or *maharlika*, who had rights to a portion of agricultural land in the *barangay* and who owed the *datu* nothing but their occasional labour. Beneath the *timawa*, there were slaves, or *alipin*, who were subordinated in a system of debt peonage (Scott, 1994). Their debt, and therefore their own allegiance, could be transferred between *datus* making them similar to bonded slaves in the European context, but like the *timawa*, they too could claim and inherit agricultural land from which they would have to pay a tribute at harvest time. Their position was not, therefore, directly equivalent to slavery in a European understanding and such distinctions were to cause confusion among the Spanish colonizers. On observing 'Filipino' society in the late sixteenth century Legazpi noted that

> [t]he inhabitants of these islands are not subjected to any law, king or lord. . . . He who owns most slaves, and the strongest, can obtain anything he pleases. . . . They recognize neither lord nor rule; and even their slaves are not under great subjection to their masters and lords, serving them only under certain conditions.
>
> (Legazpi, 1569; cited in Reid, 1988: 120)

As Reid points out, across Southeast Asia there was 'a combination of sharply stratified hierarchy with seeming looseness of political structure which would baffle European travellers, empire builders, and ethnographers for centuries' (Reid, 1988: 120).

In summary, several features of pre-colonial society are worth emphasizing because they address the 'bafflement' which European colonizers would feel, and they highlight the complex relationship of interaction and negotiation that was initiated between indigenous and outside influences (which continues into the present). Firstly, pre-colonial society was based on close familial ties and networks that defined social standing and represented the first call on personal loyalty. The origins of the *barangay* as an extended kinship grouping meant that families remained closely knit and in close proximity – a system that continued and was extended through the practice of fictive kinship. Secondly, the relatively loose system of power relations meant that allegiance was owed not to a place or an institution, but to an individual with whom a personal relationship was established. This was, moreover, a relationship between patron and client, with mutual responsibilities, not one of absolute sovereign power of one person over another. Thirdly, land ownership was not a European system of private property, but one in which usufruct rights were assigned and understood while ownership, to the extent that it was a relevant concept, remained communal. Fourthly, pre-colonial Tagalog society was characterized by a dispersed pattern of settlements with little political coherence, meaning that diverse regional identities remained powerful. Fifthly, despite the insular nature of communities, they nevertheless had trading links and familiarity with a range of other material and symbolic cultures with whom they were exchanging goods, ideas, linguistic traits and occasionally blows. Thus although ethnologically of Malay descent, and with languages of Malayo-Polynesian origin, by the sixteenth century the culture and economy of the Tagalogs and other regional groups were blends of distinctive local characteristics and the influence of outsiders. Sixthly, gender relations were decidedly at odds with those who came later. Women enjoyed considerably greater economic independence in sixteenth-century Tagalog society than was true of European societies at the time. Women were family accountants and could administer assets without their husband's consent. In general, men controlled social and sexual realms, but women exercised authority in productive and ritual domains (Eviota, 1992). Finally, religious observance was based on animistic beliefs. These practices were latterly influenced by Malay customs through contact with Borneo and supplemented or replaced by the spread of Islam, also from the south.

It was into this setting that the Spanish, variously assertive, bemused, and scandalized, inserted themselves. Yet despite the best efforts of the colonizers and missionaries over several centuries, these are also characteristics that can be seen to varying degrees in contemporary Filipino society. Beyond the symbolism and piety of Spanish Catholicism and the materialism of American-style modernization, elements of pre-Hispanic 'Filipino' culture endure in modified forms.

Spanish colonialism: galleons, governors and godliness

While 'galleons', 'governors' and 'godliness' form the popular conception of Spanish colonialism (with the addition of 'gold' in the Americas), generalizations about the Spanish colonial project in the Philippines are, in fact, hard to draw. The complexity of the Islands' experience of colonialism derives from several sources. Firstly, colonialism had its own geography as different parts of the archipelago experienced subjugation in distinct ways. From the plantation workers on Visayan sugar estates, to the tenant farmers of Luzon, to urban dwellers in Manila, to the swidden cultivators of the Cordillera mountains, the Spanish presence meant very different things. Indeed, Spanish influence during the first two hundred years of colonization was geographically highly circumscribed and the influence of colonialism beyond Manila and its hinterland waned rapidly. Secondly, the Spanish colonial project changed over time. The early boom decades of the galleon trade, the later stultifying effects of its limitations on trade, and finally the incorporation of the islands into the nineteenth-century world economy, all mean that it is possible to talk of a colonial legacy but not of a unitary colonial experience in the Philippines. Finally, there was a socially differentiated dimension to colonialism. As colonists and missionaries attempted to impose a European social structure (and morality) upon the native population, the pre-existing social structure was reworked but not replaced. Thus those with different social positions prior to colonization experienced the process in distinct ways (Rafael, 1988).

These factors make a comprehensive account of the colonial period impossible here (see Cushner, 1971, or Phelan 1959, for attempts at such an account). Instead, I will highlight several features of the Spanish colonial period: trade and production; social structure and government; and, religious conversion. Each demonstrates the complexity of the relationship between the 'inside' and 'outside' – the 'global' force of colonialism articulating with 'local' social processes among subjugated peoples. Each feature continues to exert an influence over contemporary patterns of Philippine engagement with its 'global' context.

Trade and production

The extensive trading network already established in Southeast Asia by the sixteenth century was known to the Spanish at the time of Magellan's arrival in the islands in 1521 and then Legazpi's successful conquest in 1565. Similarly, the cultural, religious and linguistic interaction between different parts of the region stretching from the Arabian Gulf to China must also have been evident. Spanish conquest was at least partly based on a desire to profit from these existing trading networks by bringing their products to European markets.

Much of Manila's wealth in the years after conquest was derived from its status as a trading port for galleons carrying Asian goods to Spain's colony in Mexico. Galleons sailed regularly between 1565 and 1813, carrying mail from Spain and Mexican silver to Manila, and Chinese merchandise, particularly textiles (from the pre-existing Manila–China trade), back to Acapulco (Caoili, 1988). Merchants in Seville and Cadiz, however, resented the competition from Chinese silk that undermined their monopoly in the Americas (Cushner, 1971). Pressure from this constituency led to tight controls over trade starting in 1593. Spanish authorities imposed a limit of one merchant fleet per year on the Manila–Acapulco route, attempted to enforce a system of quotas on the volume of trade, and restricted non-Asian Philippine trade to Mexico. This prevented the full incorporation of the colony into a growing world system and restricted its role to that of an entrepôt port on the periphery of the Spanish Empire. Local wealth accumulated through trading relationships and rental arrangements rather than value added in production – an early incarnation of the system that Yoshihara (1988) has described as 'ersatz' or rentier capitalism in the post-colonial era.

The principal beneficiaries of this system were the Spanish merchants of Manila and the Chinese traders ghettoized in Manila's Parian who, despite periodic harassment and expulsions, monopolized the retail trade and credit markets through community and clan networks. The numbers of Chinese in Manila grew rapidly in the early years of Spanish rule, as trading opportunities related to the galleon trade expanded. By 1650, Manila population comprised 7,350 Spaniards, 15,000 Chinese and 20,000 Filipinos, each segregated in distinct parts of the city. It was the Chinese who created the trading networks in the Philippine countryside that provided early Spanish settlers with commodities for export to the Americas (Steinberg, 1990). Nearly all Chinese immigrants over the subsequent centuries were men, and inter-marriage with Filipinos created a Chinese-Filipino *mestizo* group with varying levels of integration and identity with indigenous culture and society. By the nineteenth century the 'Chinese' were the country's leading entrepreneurs (see Wickberg, 1965).

Control over the galleon trade gradually became concentrated among fewer and fewer families, and the volume of trade actually declined from 1650 until the 1780s (McCoy, 1982). A more open trading regime and a sharper focus on economic development was induced by wider geopolitical events. The British occupation of Manila and Cavite in 1762–4, at the end of the Seven Years' War, concluded with onerous terms for the Spanish colony and Manila was left in economic ruin. At roughly the same time the Bourbons secured a firm grip on the Spanish throne (Cushner, 1971). They viewed colonies as sources of income for the mother country, leading to a concerted effort to develop local economic resources to the full. Numerous projects conducted by individual entrepreneurs sought to exploit resources such as pepper, clove, cinnamon, sugar cane, indigo, cotton, tobacco and timber. Incentives were also introduced to encourage agriculture, mining and silk production.

The late eighteenth century, then, was a period of economic expansion in the Philippines, at least for those few Spanish and *mestizo* merchants and landholders who were in a position to take advantage of economic opportunities. Foreign traders were permitted to operate in Manila for the first time from around 1789, and included American, British, Portuguese and French ships (Cushner, 1971). By the mid-nineteenth century considerable amounts of foreign and local capital had been invested in Philippine agriculture to supply the export trade.

The port of Manila was finally opened to free trade in 1834, by which time the extensive operations of British and other trading houses had made the restrictive colonial trade regime anachronistic. Provincial ports, such as Iloilo, were similarly opened to direct foreign trade in 1855. Throughout the nineteenth century it was British, not Spanish, capital that dominated the Philippine export economy, and

British imports and exports accounted for over half of the Philippines' trade throughout the mid-nineteenth century. Sugar in particular was a focus of British attention (Larkin, 1992).

By the mid-nineteenth century, the cultivation of cash crops had become widespread. Light industries also started to develop around Manila, which remained the colony's main port. By the late nineteenth century, the Philippines had become an exporter of primary products to the rapidly expanding world markets for sugar, hemp, tobacco, coffee, indigo and other commodities. Central Luzon became a major rice granary for both the country and the broader East Asian region, Bicol developed hemp for the American market, the Cagayan valley grew tobacco, and the Western Visayas, especially Negros, cultivated sugar. The emphasis on cash cropping, however, lead to further land concentration, indebtedness and impoverishment in rural areas, while the principal beneficiaries were foreign trading companies, urban traders or professionals, and the landed gentry. Just as early colonial rule had exempted local *datus* and *cabezas de barangay* from tribute taxes and forced labour in order to create a local comprador class, so in the nineteenth century enough natives benefited from Spanish rule to buttress Spanish authority.

The growth of Manila brought demand for agricultural products, making the provinces around the capital prime agricultural land. Land concentration was ongoing and by the eighteenth century Catholic religious orders were the largest landowners around Manila. These friar estates produced sugar, rice, fruit, tobacco and other crops, while the institution of private property meant that farmers themselves became tenants or farm labourers (Caoili, 1988). Tagalog opposition to the impositions of their friar-landlords led to periodic peasant unrest culminating in an agrarian revolt in 1745. This anti-clerical theme was also the basis of the more concerted and successful revolt based on nationalistic aspirations that would lead to independence from Spain in 1896.

Social structure and government

In the late sixteenth century, large areas of the islands were brought under Spanish control and in many places this was done without resorting to force. Instead, effective control was established through the work of missions and through the governor's application of judicious intervention to take advantage of the disunity and rivalry between native rulers (Caoili, 1988).

The Spanish did not attempt to reorder the existing social hierarchy but instead co-opted it to act as the local colonial government. Existing *datus* became the *principalia* class from which officials were chosen. The power of the few was further enhanced as the Spanish implemented a policy of *reduccion* in which dispersed settlements were amalgamated into towns and the population was forcibly resettled around the municipal hall and church. This facilitated a more intrusive form of colonial government.

Datus were eager to take advantage of the new concepts of alienable private property, title deeds and other novel legal instruments in order to enhance their wealth. Religious orders and private speculators, meanwhile, were in need of the land which the *datus* readily supplied – from their own usufruct holdings, that of their families and slaves, and from uncultivated land in the *barangays* (Scott, 1994). In addition, many communal areas simply became parts of land grants made by the Spanish crown to wealthy Spaniards, native *principales*, and religious orders.

But beneath the imposed authority of the Spanish, an antecedent system of power relations continued to operate at a local level. Dynasties descended from *datus* dominated regional political economies and in many cases had a strong vested interest in the continuation of colonial rule. Colonial rule both consolidated their power and legitimized its extension. A new hierarchy of administration was imposed by the Spanish that superseded but also incorporated traditional systems of authority. At its pinnacle was the king of Spain, followed by the Council of the Indies, viceroys and governors, local *audencias* and *alcaldes* (mayors), and finally *cabezas* (village heads). In addition, a social hierarchy organized along racial lines developed and became integrated with the administrative and economic hierarchy. That hierarchy established the level of social mobility in the colonial system. At the top were European-born Spaniards (*Peninsulares*), then Philippine-born Spaniards (*Insulares*), Spanish-Filipino *mestizos*, Chinese-Filipino *mestizos*, Chinese, and natives (*Indios*).

But the most telling aspect of the changes wrought by the Spanish was not simply the rigidification of the existing social structure according to European notions of power and subordination. More powerful still was the way in which the pre-existing social hierarchy was extended far beyond the existing level of the *datu* to reach a regional, national and global level through the municipalities, provinces, colony and the throne of Spain. The scale at which power was exercised had thus been telescoped to the global level, with local ruling elites buying into this system because of their vested interest in the consolidation of power and exemption from the duties that the colonists were in a position to insist upon – namely, corvee labour, tribute and forced purchase of agricultural goods.

Catholicism and a global spiritual hierarchy

A defining goal of Spanish colonialism in the Philippines was the conversion of the native population to Catholicism. Even Legazpi had brought a team of Augustinian missionaries for this purpose in 1565 (Caoili, 1988). But the relationship between church and state was a complex one. Under a papal edict of 1508, Pope Alexander VI had granted the Spanish throne the status of church patron and conferred responsibility for the conversion of natives in the New World. The rights of the patron to appoint bishops and secular priests was not formally a part of this agreement with the Vatican but it became standard practice. This gave the Spanish state considerable influence in the Philippines and the church was closely implicated in colonial control, particularly in remote rural areas.

The evangelization process spread outwards from Manila and extended to Cagayan in the north and Zamboanga in the South. Missionaries used native languages as the medium for teaching the church's doctrines, but limitations of vocabulary meant that Spanish words were used for key theological concepts. The result seems to have been that foreign concepts became incorporated into local understandings but only *through* those local imaginations. Thus, it appears that the native population interpreted the pillars of Catholic theology through the prism of their own worldviews and beliefs. Vicente Rafael (1988) argues that this was particularly true in the case of ideas such as conversion, submission, hierarchy and exchange which were translated by the Spanish through indigenous cultural concepts such as *hiya* (shame) and *utang na loob* (debt of gratitude). The result was that 'attempts to subordinate Tagalog idioms of reciprocity to Christian concepts were problematic and inconclusive' (Rafael, 1988: 123).

Ultimately, then, it seems that Tagalogs and others managed to circumscribe, at least in the early colonial era, the subordination that the Spanish attempted to impose. One must be suspicious at the ease with which the Spaniards accumulated both 'sovereignty' and 'converts' to Catholicism. Given the very different understanding of power, authority and spirituality among native people, it seems likely that the easiness of the task reflected the fact that the Spaniards' requirements for pledges of allegiance and faith meant little to the local people. Rafael provides a convincing argument that this was because native notions of a debt of gratitude were based on an ongoing relationship of indebtedness in which the debt is never fully repaid, for example to one's mother. The whole concept of power and subordination was therefore different. Similarly, Rafael points out that the word *tauad* (or *tawad* – to bargain, haggle or evade) represents the Tagalog translation of confession or pleading for forgiveness. Clearly the implication is that one bargains in a two-way process with the figure of authority (ultimately, God) for forgiveness and salvation, a notion that would have scandalized the Spanish friars. Several accounts also suggest that worship of ancient gods and the continuation of pagan festivals and rituals was widespread for several centuries after conquest (Cushner, 1971). Even in the late twentieth century, Catholicism has not eradicated apparently incompatible beliefs in witchcraft, spirits and the power of talismans (Lieban, 1967).

Like the secular government, the Spanish clergy imposed a particular conception of social hierarchy. Through church administration and theological doctrine, the laity learned that authority resided elsewhere. Starting with local priests the hierarchy spread upwards to bishops and the archdiocese of Manila but ended with the Spanish crown and the pope. An entirely new and 'globalized' hierarchy was imposed upon the native population. In its most insidious form, this hierarchy was evident in the use of language. Rafael notes that while native tongues were used by missionaries, Spanish was retained as an elite language not to be used by locals. Throughout the colonial period, although Spanish words were used in Filipino dialects – most significantly for the holiest of religious concepts that missionaries did not feel could be translated 'downwards' – Spanish remained the language of the elite. Beyond Spanish there was Latin, the language of the senior clergy and learned laity.

Thus in both church administration and in cultural translation, an implicit hierarchy was established that placed native ways of life at the bottom and privileged those brought from, and dictated by, the outside. It would perhaps be overstating the point if this feature of colonialism were too directly linked with contemporary deferral to, and privileging of, Western culture, but it seems that at least some of the current authority that a globalization discourse draws upon is provided by the sense of hierarchy established through linguistic, moral and religious aspects of colonialism.

In summary, several points emerge concerning the legacy of Spanish colonialism in the Philippines. Firstly, the experience tightly restricted economic expansion and when resource development did occur it was conducted by European and North American trading houses investing in export commodities. The result was a lasting legacy of dependence on selling primary products in volatile world markets. Secondly, colonialism led to the formation of an elite social class with a vested interest in continued outside involvement. Existing hierarchies were both rigidified and extended to form a social structure that reached ultimately to the Spanish crown as the font of power. Thirdly, the colonial experience produced a profound and yet partial absorption of European practices and beliefs. Finally, the Spanish attempted to establish a cultural hierarchy privileging the outside and foreign. This was evident in religious, racial, and linguistic relations.

America's colonial experiment

While Spanish colonialism lingers as a legacy, American imperialism, even five decades after independence, still reverberates through Filipino life (see Plate 1). Where the Spanish had retained their language for themselves and for the local elites, the Americans set about developing a comprehensive countrywide education system, and establishing English as the lingua franca of their new territory. American rule likewise left its mark on other areas of public service provision, notably government and healthcare. Three main themes in American colonialism are germane to this chapter: the entrenching of economic dependence; the establishment of new political and social structures; and, the reworking of cultural understandings through the educational system.

American colonialism and economic dependence

The early American regime in the Philippines was keen to divert trade in commodities to the US market, but showed little interest in reshaping the basic structure of its colony's economy (Owen, 1971). Indeed US tariff policy deepened the dependence of the Philippine economy on a small number of agricultural exports. The Payne–Aldrich Tariff Act of 1909 and the Underwood–Simmons Tariff Act of 1913 led to virtually free trade with the continental United States. American manufactured goods were imported duty free and Philippine commodities such as sugar, abaca and coconut oil passed freely into the growing US market. Tariffs were, however, applied to Philippine manufactured exports with more than 20 per cent non-Philippine content.

The depression of the 1920s and 1930s changed the economic context of US

Plate 1 Postcolonial complexity: all-American dolls dressed in Spanish costumes on sale in 1998 to celebrate the centenary of the Philippine revolution against colonial rule

relations with its colony. Quotas were imposed on agricultural goods such as sugar, cigars, cordage and coconut oil through the Tydings–McDuffie Act of 1934. Beyond these quotas, import duties would be charged. The duty-free quotas were intended to be phased out with the move towards political independence, but in practice lasted until 1974.

It is important to recognize that these American policies towards the Philippines were a complex mixture of vested interests and an ingenuous belief that they represented the best interests and will of the Filipino people. The vested interests at work were the lobbies representing American manufacturers and agricultural producers who favoured free export of goods and restricted importation of products respectively, and the Filipino landholding elite who had, by the end of the Spanish period, amassed considerable wealth. The interest of the Filipino elite was in continued access to the immensely profitable US market and these views were heard by the US authorities. They were joined by a growing lobby of American business interests in the Philippines. Given the almost non-existent state of Philippine export manufacturing at that time there was no constituency calling for greater protection from imported manufactured goods. Nor was there any politically powerful voice calling for a more equitable distribution of land holdings. The American ideal of the manifest destiny of frontier expansion and a benign civilizing mission might have underpinned the colonial experiment, but it certainly did not extend to ensuring that the Jeffersonian model of small independent farmers replaced the tenancy model inherited from the Spanish. Even in the case of the appropriated religious estates in Cavite, little was done to prevent them falling into the hands of already powerful local families (Endriga, 1970).

What emerges then, is a picture of the American regime motivated by both self-interest and altruism. Self-interest in ensuring a supply of agricultural commodities and dumping manufactured goods in the Philippines, and altruism in the belief that in doing so they were acting benignly and according to the wishes of 'the Filipino people.' But the Filipino people with a voice were exactly those with a vested interest in the arrangement that developed. Little wonder, then, that many in the Filipino elite saw no particular advantages in seeking independence from the USA and letting go of the 'special relationship' was a slow process.

The result was that ultimately, US rule had little structural effect on the economy of the Philippines. In the 1890s, as in the 1930s and the 1950s, the economy was characterized by '[o]verdependence on a few exports, tenantry, indebtedness, low productivity, corruption and inefficiency, undercapitalization, [and] miserable working conditions' (Owen, 1971: 55). In 1946, political independence was granted to a country firmly dependent on agricultural exports, with just four crops accounting for nearly 90 per cent of all Philippine exports (as they had in 1894). The profits from such exports – and it was an immensely lucrative trade – were concentrated among a small number of elite families.

Colonial government and society

A second feature of US colonialism that yielded greater change than economic policy was the nature of political and social organization that was installed. Like the Spanish, the Americans did not neglect to notice that a class structure already existed and they recruited government officials from its upper echelons. The result was that the voices heard by the Americans were those of the elite. This does not necessarily exonerate the US regime from responsibility for the continuing extreme poverty in many parts of the islands, but it goes some way towards explaining why a regime ostensibly based on high-minded ideals could pursue policies so contrary to the interests of so many.

In the early years of colonial rule local elections were initiated, in line with the 'enlightened' American mission to train their Filipino counterparts for self-government. But the institution of elected office was grafted onto a circumstance of highly unequal economic and social power relations and an authoritarian political culture. The result was that local democracy served solely to entrench the power of economic elites in the political sphere (Cullinane, 1971; Doronila, 1992). Local political dynasties were able to perpetuate their control with the added legitimacy afforded by 'democratic' principles. Just as had occurred under Spanish rule, local elites were able to use the influence, ideas and institutions of outsiders to entrench their own power.

While enfranchising, at least in theory, rural populations, the American political system also made each level of government dependent on higher ones (Lopez, 1966). Just as the Spanish had extended the social and religious hierarchy, so the Americans extended the political pyramid from villages, to municipalities, to provinces, to Manila, and ultimately to Washington. Improvements in communications and transport infrastructure made centralized decision-making feasible and local politicians operated in this system through the centuries-old system of patron–client relationships.

Teaching modernity

Perhaps the most profound impact of American rule in the Philippines was exercised through the educational system. Golay describes the American-staffed public school system as 'a tool for communicating the idea of change to the grass roots of Philippine society . . . instrumental in intensifying the Western identification of Filipinos who had been bypassed by the Spanish cultural impact' (Golay, 1961: 409). School history and culture texts, for example, invariably had (and still largely have) a 'they-taught-we-learned' presentation in which the Filipino subject 'stands naked and in need of being dressed in foreign gear' (Mulder, 1990: 91). The nationalist historian Renato Constantino goes further and suggests that education was a key tool of American colonialism. In a 1966 article entitled 'The Miseducation of the Filipino' he argues that the purpose of the public school system was to train Filipinos to be good colonial subjects, conforming to American ideals:

> The new Filipino generation learned of the lives of American heroes, sang American songs, and dreamt of snow and Santa Claus. The nationalist resistance leaders . . . were regarded as brigands and outlaws. The lives of Philippine heroes were taught but their nationalist teachings were glossed over.
>
> (Constantino, 1982: 6)

According to Constantino, the impact of this sort of cultural indoctrination was also felt in the economic sphere. He argues that a generation grew up with a perception of their country based on bucolic rural imagery, for example in the landscape paintings of Amorsolo. This imbued a belief that the Philippines was essentially an agricultural country and destined to remain so. Consumption habits, meanwhile, became oriented towards American manufactured products and the implication of foreign superiority was subtly inculcated: 'our books pictured the Western nations as peopled by superior beings because they were capable of manufacturing things that we never thought we were capable of producing' (Constantino, 1982: 8).

In summary, the results of American colonialism were the continued dependence of the Philippine economy on agricultural exports and its domination by an elite landholding oligarchy that had enjoyed a deepening of their control and its legitimation through democratic processes. Thus, US power fed into *local* networks of power. More insidiously, US rule saw a cultural orientation towards the West in identifying modernization as desirable and in defining its direction. The centralization of political power and the establishment of a public school system, notwithstanding the high ideals of many involved, played an important role in colonizing consciousness. What is now popularly referred to as 'colonial mentality' in the Philippines, together with the difficulties in fostering a nationalist vision of development, can perhaps be attributed in part to this legacy.

The post-colonial state, 1946–66

Philippine independence in 1946 represented the culmination of a planned transfer of sovereignty starting in 1935 with a Commonwealth government, but interrupted by a destructive wartime occupation by Japanese forces (1942–5). In the years after 1946, as in the years before, American commitment to political self-determination for the Philippines was notably higher than its vigour in ensuring economic autonomy for the former colony (Doronila, 1992: 19). Perhaps because independence was achieved without the 'clean break' of a revolutionary struggle, American influence continued to be exercised in a variety of spheres.

Studies of US involvement in the Philippine political economy provide detailed accounts of the post-war period (for example, Boyce, 1993; Cullather, 1994; Doronila, 1992; Hawes, 1987). Rather than attempting to provide a comprehensive summary of this literature here, I will use the case of trade and industrial policy to highlight the continuation of a negotiated relationship between the Philippine political economy and outside influences in the period 1946–66.

Independence had been legislated by the Tydings–McDuffie Act of 1934 in the US Congress, and this allowed for a ten-year transition in which preferential trading arrangements would end on 4 July 1946. Wartime destruction and the decimation of the Philippine economy, however, led to a reassessment of the situation in the form of the Bell Trade Act of 1946 which was to establish the trading relationship between the two countries for the subsequent twenty-eight years. It dictated free trade until 1954, followed by a gradual increase in tariffs until 1973, when full duties would be imposed. In addition, the Philippines could not impose taxes on exports to the USA, absolute quotas were established on seven important Philippine commodity exports, and the peso exchange rate was fixed to the dollar. Finally, the Act required the Philippines to extend parity rights to US citizens in resource exploitation activities and public utilities (Doronila, 1992). The result of these conditions was to establish a post-war economy that retained its dependence on agricultural commodity export, to continue the dominance of Philippine politics by the landed oligarchy, to ensure US control over significant areas of the economy, and, finally, to allow the USA some powerful economic sanctions with which to exercise its influence as political and military 'patron'. It was in this capacity that the USA was able to insist on a Military Bases Agreement in 1947 that allowed a continued military presence in the islands.

Given the terms of trade under which the country operated, it was inevitable that a foreign exchange crisis would be precipitated eventually. In 1948, the Philippine government persuaded Harry Truman to assent to exchange and import controls in the form of Republic Act 330, The Import Control Act. Controls were intensified during the 1950s to become an instrument for an import substitution industrialization (ISI) programme (Doronila, 1992). This was less an example of altruism on the part of the USA than a reflection of the fragility of the new Philippine state which the Americans worried would collapse and fall into the hands of the communist *Huk* rebels who were then riding on a wave of rural unrest (Kerkvliet, 1977). Just months before, tumultuous events in China had jolted American policy makers, and they hoped that the Philippines would serve as an example of 'moderate nationalism' in opposition to communism (Doronila, 1992). In achieving some level of economic autonomy, then, the Philippine state was able to play on wider geopolitical concerns.

In addition to quota limits on imports, new import substitution industries enjoyed tax exemptions, liberal credit facilities, and windfall profits due to the over-valued pesos, but the programme also provided opportunities for political patronage in the allocation of foreign exchange licences and incentives. A predominantly agricultural economy became reoriented towards packaging, assembly and light manufacturing, with the share of manufacturing in net GDP rising from 10.7 per cent in 1948 to 17.9 per cent in 1960 (Doronila, 1992: 55).

The dominance of export producers, such as sugar planters, was clearly compromised by the ISI programme and several writers frame this period as one of diversification in the Philippine social structure as it divided along sectoral and ethnic lines (Hawes, 1987; Rivera, 1994; Yoshihara, 1985). The economic interests of the elite landowning class, mostly Spanish creoles or Chinese *mestizos*, became widened to include entrepreneurial manufacturing in the import substitution sector. But the ISI sector also included two other domestic groups. With the passing of the Retail Trade Nationalization Act in 1953, Chinese entrepreneurs were banned from a sector they had dominated and began to move their capital into manufacturing enterprises. Indigenous Filipinos, meanwhile, were also entering the manufacturing sector and taking advantage of preferential access to government resources and privileges (Rivera, 1994; Yoshihara, 1985). Foreign interests, notably American, also entered the domestic manufacturing sector to take advantage of favourable economic opportunities. These interests, benefiting from production behind the new trade barriers, moved away from a political orientation in favour of free trade. At the same time as it satisfied these economic constituencies, the ISI programme also went some way towards meeting the demands of a growing nationalist movement in the country, led by politicians and intellectuals such as Claro Recto and Jose Diokno. There was thus both a political and a transnational economic coalition of interests behind import substitution.

By the late 1950s, however, a slowdown in growth occurred as the marginal returns from local assembly of 'knocked-down' manufactured goods began to diminish. A balance of payments crisis also emerged due to the import of machinery, raw materials, tools, parts etc. by ISI industries (Ofreneo, 1995). These circumstances started to increase pressure on the government to move away from trade and exchange controls. But the national and transnational coalition supporting ISI had already grown politically powerful. Thus when decontrol was initiated in 1962 by newly elected President Macapagal, it was a piecemeal process that represented a compromise between interests. Macapagal had stood on an electoral platform of free enterprise and was under political pressure from the USA via the IMF, which was responding to the interests of its export sector. As a result of this external pressure, the peso was devalued and the system of exchange controls was dismantled, but Macapagal was able to reassure the business community that they would still be afforded some protection by a tariff system (Doronila, 1992: 66).

Over the subsequent decade of prevarication in the 1960s, a class of professional technocrats emerged with increasing influence. Most notably, the conservative economist Gerardo Sicat was developing a critique of protectionist policies and arguing for the depression of wages to promote industrial expansion, rather than policies of import control. Sicat was later to become an influential member of President Marcos' economic staff and his ideas were largely implemented in the 1970s (Doronila, 1992). Technocrats such as Sicat were not, however, gaining increasing influence simply through the merits of their arguments. During the 1960s the political climate was changing. The oligarchy that had dominated the political and economic life of the country was starting to experience a 'breakdown of cohesion' (Hawes, 1987: 36). In addition, the Philippines' foreign exchange crisis in the late 1950s had led to IMF involvement in the economy in the form of policy prescriptions and financial assistance, although first hand accounts suggest that the IMF agenda was being driven largely by US interests (Doronila, 1992: 114). Nevertheless, the shift from bilateral relations with the USA to dealing with the multilateral international financial institutions was a significant change in the Philippine political context.

The rising fortunes of Filipino technocrats were one result of this changing context, but despite their preference for an export oriented strategy of development, economic policy during the 1960s remained essentially based on import substitution. Although the peso was devalued and exchange controls were relaxed, tariffs were increased, and tax exemptions and cheap credit were still provided to domestic industry. A significant lesson to be drawn from this is that competing policy prescriptions were played out through the political culture of the Philippine state in a distinctive way. Macapagal was able to garner US and IMF support by devaluing the peso, and yet at the same time, other political pressures meant the retention of many elements of the ISI programme.

The Marcos regime and the world economy, 1966–86

Eventually, this balancing act proved impossible to sustain and a deteriorating economy and further balance of payments crisis saw deepening involvement from the international financial institutions, particularly the IMF. In addition, a growing constituency of US and Filipino export producers in the Philippines was looking for a more favourable business climate. Thus, under Ferdinand Marcos in the late 1960s legislation started to appear that began the process of constructing an export oriented industrial (EOI) development programme: the Investment Incentives Act of 1967 (RA 5186), establishing the Board of Investments and giving a broad range of tax incentives to export producers including tax credits on raw materials and imported capital equipment; the Export Incentives Act of 1970 (RA 6135), extended these benefits with a ten year tax holiday on most materials and capital goods used in manufacturing and processing; and the Foreign Business Relations Act of 1970 (RA 5455), that removed restrictions on the repatriation of profits (Ofreneo, 1995).

Martial law

The decisive factor, however, in establishing an export-oriented industrial policy in the early 1970s was the declaration of martial law by Marcos in 1972. The abolition of Congress, and the suppression of labour organizations and dissident intellectuals enabled the President to give his technocrats a free hand in reorienting the economy. But Marcos was careful, even in declaring martial law, to ensure the support of the US government and the IMF. Numerous accounts indicate that he first sought, and received, assurances that such an action would receive no condemnation or retribution from his allies in Washington (Bonner, 1987). In fact, martial law was received with considerable enthusiasm in some quarters. The American Chamber of Commerce in the Philippines sent Marcos the following telegram a few days after martial law was declared:

> The American Chamber of Commerce wishes you every success in your endeavours to restore peace and order, business confidence, economic growth and the well being of the Filipino people and nation. We assure you of our confidence and co-operation in achieving these objectives. We are communicating these feelings to our associates and affiliates in the United States.
>
> (Reproduced in Schirmer and Shalom, 1987: 230)

Under martial law, Philippine development policy became increasingly oriented towards export production. With Marcos fully in control, and with his team of technocrats able to implement their strategies unhindered for the first time, Presidential Decree 66 established the Export Processing Zones Authority to oversee an existing zone in Bataan and to develop other zones elsewhere in the country. The Mactan EPZ, near Cebu City in the Visayas, was designated at the request of the local government and established in 1978 (Guerrero *et al.*, 1987; Chant and McIlwaine, 1995). The Baguio City EPZ in northern Luzon followed in 1979 (PD 1825), but the impetus in this case came largely from the US semiconductor manufacturer Texas Instruments. The Cavite EPZ, just to the south of Manila was designated in 1980 under Presidential Decrees 1980 and 2017, and construction started in 1981 (McAndrew, 1994). By locating in these zones, firms were eligible to benefit from a variety of financial incentives, simplified regulatory frameworks, and established infrastructure and services.

In addition to fiscal and regulatory incentives aimed at foreign investment and export promotion, the Marcos government also embarked upon aggressive marketing campaigns to represent the country as a desirable node in the global matrix of travel and investment opportunities. A few months after martial law was declared in 1972, Fortune magazine carried the following advertisement placed by the Philippine government:

> To attract companies . . . like yours . . . we have razed mountains, felled jungles, filled swamps, moved rivers, relocated towns, and in their place built power plants, dams, roads . . . an executive centre and a luxury hotel. All to make it easier for you and your business to do business here. And we've done more. Much more.
>
> (*Fortune Magazine*, 12 October 1972)

At the same time the government, largely through the activities of Imelda Marcos in her capacity as Governor of Metro Manila, pushed through a variety of urban megaprojects in a successful attempt to attract high-profile world events, such as the 1974 Miss Universe beauty pageant and the 1976 IMF/World Bank conference. The purpose was to establish Manila as a 'world-class' city. Most infamous among these projects were the concrete modernist monuments constructed on reclaimed land in Manila Bay: the Philippine International Convention Centre, the Folk Arts Theatre, the National Film Centre and the Cultural Centre of the Philippines (Pinches, 1994).

The more concerted focus of the Marcos regime, however, was on the export of primary commodities through monopoly marketing boards (Hawes, 1987). Establishing state corporations to export coconuts, sugar and fruit products, satisfied several goals: firstly, export conglomerates were intended to play a role similar to that of the industrial *chaebols* in Korea by participating in a diverse range of activities but with commodity export as their base; secondly, by taking a firm grip on the agricultural export sector, Marcos was able to control the source of wealth on which many of his opponents in the landed oligarchy depended (in addition, the suspension of the Congress, which they had dominated, effectively curtailed their formal political influence); finally, control over the export conglomerates provided lucrative patronage appointments for loyal supporters.

The export economy was not, however, limited to commodities. Starting in the 1970s, the export of human labour became an increasingly important part of the Philippines' external economy. Labour export from the Philippines started in the early twentieth century as Filipino workers were employed in the fruit plantations of Hawaii and California (Gonzalez, 1997). The earliest legislative recognition of this phenomenon was Act 2486 of 1915 passed by the US-supervised Philippine Legislature to regulate the activities of labour recruiters. It was not until the Labour Code of 1974 (Presidential Decree 442) that further legislation to address the issue was adopted. By then, the devalued peso and domestic economic stagnation left large numbers in disguised unemployment in the agricultural sector and made a 'dollar' income highly attractive. These circumstances coincided with 'pull' factors abroad:

the oil boom in the Middle East that provided opportunities for men to work in construction and other occupations; the growth of containerized shipping provided openings for seamen; and, the rapid growth of East Asian economies (especially Hong Kong, Singapore and Japan) created wage differentials that made domestic work financially appealing even for college educated Filipinas. The export of labour was actively encouraged by the Marcos administration which created various agencies to oversee the recruitment, placement, licensing, foreign exchange remittances, dispute resolution, registration and documentation of overseas contract workers (Gonzalez, 1997).

The politics of economic strategy

The politics of this period are complex and the Philippine economy was liberalized and opened to the world system in only selective ways. Despite an avowed belief in export promotion and economic liberalization, the 1970s also saw continued adherence to some aspects of import substitution strategies due to local political expediencies (Ofreneo, 1995). Examples included regulations concerning the nationality of ownership in particular sectors of the economy and tariff protection for certain goods (see Hawes, 1987).

Several different themes have been emphasized in coming to terms with the forces at work in the Philippine political economy at that time. Some choose to portray the period in terms of an emergent transnational technocracy in support of an export-oriented industrialization strategy. Thus there was a domestic, though often foreign trained, technocratic corps whose thinking reflected that of the IMF and World Bank. The constant need for support from these institutions meant that such policies could be forced upon the Philippine government – or, more accurately, the constituency within the government in favour of such policies could be strengthened (Broad, 1988).

Another point of view would emphasize the role of private economic interests in shaping policy. Domestic and foreign export manufacturers clearly had a strong interest in the sorts of incentives being offered, and producers of primary export commodities favoured a weak peso. Both sectors wished to see a close linking of the domestic economy with international capital flows, but at the same time, the domestic 'merchant class' retained certain elements of a self-interested protectionist position (Koppel, 1990).

Yet another approach is to view the emergence of an EOI regime in terms of the personalized brand of political economy that became predominant during Marcos' martial law regime. Modelled after Korean *chaebols* or Japanese *zaibatsu*, Marcos attempted to create agro-industrial conglomerates that would lead the way for Philippine products in the global economy (Hawes, 1987; Ofreneo, 1980). Heading these organizations were close friends and relatives of the President.

Supporting the economy, and disguising the system's inefficiencies and corruption, was the flow of money coming from the IMF. This source of outside funding enabled the President and his 'cronies' to insulate themselves effectively from the opposition of those families who were not within his circle of influence and patronage. Added to this was the continued military, political and financial support being provided by the United States. But it is important to note the subtlety with which Marcos managed these outside supporters. While US influence was considerable, Marcos could play a power game in which he emphasized the threat of communist insurgency and played on American fears of losing a non-communist foothold in the region. At the same time, the IMF continued, in a sense, to fool itself that Marcos was implementing the sort of reforms it wanted through the good offices of like-minded technocrats. Marcos was indeed employing a rhetoric that was pleasing to international financial institutions, but in practice the economy was far from an undistorted free market. It was dominated by monopoly marketing boards for export commodities and 'cronies' eliciting preferential treatment from the government. Gary Hawes emphasizes the highly political nature of Marcos' rule:

> Ferdinand Marcos, unlike the international actors who supported the Philippine state, was always clear that his interests were not completely synonymous with those of the multinationals, or the US government, or the World Bank/IMF group. His primary goal was to remain in office. . . . As long as Marcos remained president, he was able to use the coercive and administrative organizations of the state to his own end. He threatened, bluffed and took action whenever possible to see that, while he followed the prescribed path to development, while he enlarged the role that foreign investors could play, he did nothing to endanger his own continued rule.
> (Hawes, 1987: 152–3)

Marcos' politicization of the economy also provided a tool with which to deal blows to his political enemies. Some of the wealthiest families in the country, most notably the Lopez clan of Negros, were ostracized on account of their political rivalry with Marcos (McCoy, 1994b). Many took refuge abroad, but those who stayed were prevented from taking a leading role in investing in domestic industry. Marcos could do this because foreign rather than domestic capital was the major source of investment generation in the country, allowing him some insulation from the landed oligarchs who had dominated the economy to that point. The result, however, of the foreign capital influx in the 1970s, particularly from multilateral agencies, was the accumulation of massive debts (Koppel, 1990).

Economic decline and the fall of Marcos

The unviability of this debt driven economic structure meant that the government was continually dependent on flows of credit from the IMF and the World Bank. As a result, the policy prescriptions of these institutions became more deeply entrenched, especially as the World Bank started to attach conditions to its loans under the Structural Adjustment Program (SAP) established in the late 1970s (Bello *et al.*, 1982; Broad, 1988). The SAP dealt some heavy blows to domestic industry at a time when worldwide recession was also undermining the markets for export commodities. The result was a deep recession in the Philippines and growing social unrest. The assassination of the opposition leader, Benigno Aquino, in 1983 added to the sense of political crisis and eventually resulted in the 'EDSA' revolution of February 1986 in which his widow, Corazon Aquino, was swept to power. Even in his final days in office Marcos was playing on US relations and employing his international connections to attempt to stay in office. His final departure was in a US Navy helicopter.

It hardly needs emphasizing that the Marcos regime, particularly in its latter years, was disastrous for the Philippine economy. The 1970s and early 1980s were a period of relative economic stagnation for the country. GDP growth followed a downward trend between the mid-1970s and the mid-1980s, including dramatic contractions in 1984 and 1985 of about 8 per cent in each successive year. These figures were further compounded by a consistently high rate of population growth. If the authoritarian years are compared with those preceding them, GNP per capita per year exhibited a 3.1 per cent expansion in 1962–74, but only 1.0 per cent over the subsequent twelve years (Boyce, 1993: 23). Estimates of the proportion of the population below the poverty line suggest an incidence of 43.8 per cent in 1971 increasing to 58.9 per cent by 1985 (Boyce, 1993: 46).

International confidence in the Marcos regime gradually ebbed following the 1983 Aquino assassination and the realization of a deep crisis of foreign indebtedness. This period was also one of declining trade, with the value of exports almost unchanged between 1980 and 1986.

Actors on a Philippine stage

Before discussing the post-Marcos political economy of the Philippines, some general points can be made concerning the ways in which the preceding discussion connects which the broader themes raised in this chapter. These points apply explicitly to the Marcos regime, but as Doronila shows, the tendencies which Marcos took to extremes were already well established in the post-independence Philippine political economy (Doronila, 1992).

The first point concerns the balance of power between internal and external forces in determining economic policy. Foreign influences have always represented one set of actors playing roles of varying importance on a Philippine stage, but those directing the production remain domestic players. It is misleading to portray this relationship, as has often been done, as entirely based on 'neoimperialism' by the US and multilateral banks; their role has always been mediated by, and dependent upon, a certain constituency of technocrats and vested interests within the Philippines. Equally, however, it would be incorrect to imply that these institutions were in any sense impotent in their dealings with domestic politics. Marcos was a powerful phenomenon and while sometimes impervious to the wishes of his international supporters, he was ultimately dependent on them. These supporters provided economic aid to 'insulate' the regime from domestic civil society, but also lent ideological support, legitimacy, prestige and military aid to the Philippine government. The relationship between the Philippine state and its international context is therefore a complex dialectic of agency and dependency.

A second point that emerges is the highly personalized nature of politics in the Philippines, even at the highest levels. As Rivera (1994) points out, little can be understood of the post-war Philippine political economy without reference to the web of connections that joins certain groups of families and divides others. In attempting to characterize the Philippine political economy, many have employed the 'patron–client' metaphor to indicate the nature of cultural understandings that underpin personal loyalties and a 'moral economy' of political largesse (Hollnsteiner, 1963). Recently, the notion of 'bossism' has been suggested as a more accurate portrayal of the realities of political-economic power at provincial and national scales (McCoy, 1994a; Sidel, 1995).

Thirdly, various overlapping interest groups form the cast of actors who compete to define political priorities. Classifications are numerous but they include: Chinese capitalists / ISI bourgeoisie / local landlord class / foreign capitalists (Rivera, 1994); mercantile state / merchant capital / land-owning class / international capital (Koppel, 1990); domestic Chinese capital / Filipino capital / foreign capital (Yoshihara, 1985); state capitalists / crony capitalists / domestic market bourgeoisie / export market bourgeoisie (Hawes, 1987). Each of these categories might be broken down further, and others might be added (for example, Filipino technocrats from the 1960s onwards), but the important point to note is the cultural, ethnic and political complexity of the Philippine economy. Each change in the direction of development strategy has been derived from struggles within and between these groups as they try to define their best interest and attempt to ensure that it is acted upon. The outcome of these struggles is thus a mixture of structural conditions, institutional constraints, economic sociology and individual agency. Over the last ten years the result of such a mix of factors has been the formation of a solid coalition espousing the importance and even inevitability of orienting the state's development strategy towards attracting investment and promoting export-oriented development.

The post-authoritarian years: Aquino and Ramos

In retrospect, the lasting achievement of the Aquino government was to re-establish democratic processes in the Philippine political system and to go some way towards restoring international confidence in the government. But the 'People Power' revolution of 1986 did not produce the sort of social justice agenda for which many of its participants had hoped. Instead, many of the same figures continued in power, and in a sense the events following the 'EDSA' uprising represented a reversion to the old system of landed oligarchs that Marcos had gone some way towards undermining (Anderson, 1988). Elections for local governments and Congress in 1988 returned most of the Marcos era *caciques* to power and Congress was overwhelmingly dominated by landed elites (Guttierez, 1993). Despite distortions in the democratic process, Aquino maintained a fervent faith in its sanctity and so failed to use her extensive administrative powers to act on social reform in 1986–8. Thus, for example, agrarian reform was deferred to a Congress dominated by landowners and its redistributive component was comprehensively undermined (Riedinger, 1995).

Aquino's government was constantly under threat from a restless army and survived numerous coup attempts, while trying to contain the insurgency movement organized by the New People's Army (NPA). These military threats, combined with economic fragility meant continued reliance on political and economic support from the USA and multilateral institutions. The government therefore professed a neoliberal economic framework of faithful debt servicing, reduced expenditure, deregulation and export-oriented development.

The imperatives of attracting foreign investment and encouraging export production translated into a variety of policy initiatives under the Aquino administration (Ferreria *et al.*, 1993). In 1987, the Omnibus Investments Code (Executive Order 226) reworked the financial and regulatory incentives for those locating manufacturing activities, regional headquarters or warehousing facilities in the Philippines. The code established the current framework in which foreign and domestic investments in 'priority' and export sectors are provided with incentives by the Board of Investments or the Export Processing Zones Authority. The Foreign Investments Act (Republic Act 7042) of 1991 extended regulatory leeway for investors by allowing total foreign ownership of companies except in a few strategic areas. The Act also reduces regulatory control over foreign enterprises that are not receiving incentives, allowing them to be wholly foreign owned, and permits export enterprises receiving incentives to sell up to 40 per cent of production in the domestic market.

A further component of the government's investment incentive framework is the 'Build-Operate-Transfer' (BOT) scheme of 1993 (Republic Act 7718). Under the scheme, a contractor constructs, operates and maintains an infrastructure facility for an agreed period of time during which they may charge user fees to recover investment and operating expenses. The facility is then transferred to government ownership. Major projects such as power plants, roads and Manila's Light Rail Transit system have been financed in this way.

The results of these policies under the Aquino administration were mixed. Economic indicators for the first two years of her Presidency were impressive. The removal of Marcos in 1986 brought a wave of new investment and international assistance to the Philippines over the subsequent few years. In 1989–92, however, political instability caused by recurrent coup attempts, communist insurgency, and Muslim unrest in Mindanao, combined with a global depression in economic activity, led to a slowdown in Philippine growth. Having expanded to record levels, foreign investment declined between 1990 and 1992, exports and imports levelled off, and GDP actually declined in 1991.

Ramos: liberalization and political stability

The Ramos government (1992–8) maintained the Aquino administration's commitment to export-led growth and, more importantly, achieved considerable success in establishing political stability through closer control over the military, an amnesty for rebels and negotiated peace with the NPA, and settlements with Muslim secessionists in Mindanao. The government also moved in several ways to extend the incentives offered to foreign investors and exporters. The Export Development Act of 1994 (RA 7844), styled as the 'Magna Carta for Exporters,' was implemented from the beginning of 1995. In a preamble, the Act dictates that '[t]he State shall instill in the Filipino people that exporting is not just a sectoral concern but the key to national survival and the means through which the economic goals of increased employment and enhanced incomes can most expeditiously be achieved.'[1] The Act reduces the proportion of an 'export' firm's output that must be exported to 50 per cent, thus allowing more companies to claim the set of incentives offered. These incentives include duty free importation of capital goods and tax credits for imported raw materials and inputs and to reward increases in export revenues. Deregulation and liberalization policies in the banking, oil and retail sectors form further components of the government's investment promotion agenda, all of which have drawn praise from multilateral financial institutions and creditors.[2]

These various programmes were coordinated through the government's Medium Term Philippine Development Plan, 1993–8, (MTPDP) but the galvanizing slogan for the programme of reforms was the President's 'Philippines 2000' vision – the ambitious goal of becoming a 'Newly Industrialized Country' (NIC) by the year 2000. The plan contained many laudable goals – including a stated commitment to 'people empowerment' along with achieving 'global competitiveness' (NEDA, 1995a). But as Alex Magno (1993) points out, the most powerful characteristic of the plan and the vision is as an *icon*. Like other symbolic political gestures – Mahathir's 'Vision 2020', Suharto's 'New Order' – Philippines 2000 captured an optimistic mood and placed the government on the side of positive thinking and the continually deferred dream of better days ahead. The result is that opposition is cast in the role of negativism, defeatism and pessimism (Magno, 1993). The government has even developed its own model of Filipino subjectivity in support of the MTPDP and 'Philippines 2000':

> The efforts of communicating the MTPDP/Philippines 2000 have been pushed into a more personal but dynamic dimension through the conceptualization of a modern Filipino role model – Juan Kaunlaran ['Johnny Progress']. As conceived, Juan Kaunlaran is the modern Filipino, empowered and globally competitive, who has risen above the self-deprecating images of the indolent Juan Tamad ['Johnny Lazy'] and the submissive Juan de la Cruz.
>
> (NEDA, 1995a: 47)

The MTPDP placed a heavy emphasis on global competitiveness, a theme that the President has repeatedly propagated. In his 1995 State of the Nation address, for example, Ramos declared that 'we must press on with deregulation and liberalization and bring down the last of our self-imposed barriers to economic growth left over from the age of protectionism'.[3] Again in his 1996 *Ulat sa Bayan* (Letter to the Nation) address he emphasizes the point: 'There is a new reality that underscores our national life. We are part of a new global economy – in which every nation must compete, if it is to prosper.'[4] By the end of the Ramos presidency, the wisdom of this development orthodoxy was firmly entrenched among conservative Philippine economists and government officials (Balisacan, 1994; Habito, 1993).

In addition to these new programmes, the four established export processing zones continue to offer financial benefits, regulatory incentives, land and utilities to export producers but they have been supplemented in the last few years by a variety of Special Economic Zones and private Industrial Estates that share some of the same privileges. In 1995 the Export Processing Zones Authority was reconstituted as the Philippine Economic Zones Authority to reflect this broadening mandate (RA 7916).

A variety of regional centres and growth areas have also been incorporated into the national government's industrial promotion strategy. The Subic Bay Metropolitan Authority, for example, has created a significant centre for investment using the infrastructure from the Subic Bay US naval base that closed in 1991. Other more dispersed growth corridors/areas have been established with integrated planning frameworks to foster industrial growth. An example is the East Asian Growth Area (EAGA), incorporating Brunei, Indonesia, Malaysia and the Philippines. The EAGA adopts the 'growth polygon' model to bring together capital and other factors of production in a free trade zone, in this case focused on Davao City on Mindanao (Turner, 1995). Another example is the Calabarzon area (see map 2 in chapter 3), incorporating the provinces of Cavite, Laguna, Batangas, Rizal and Quezon, which has a coordinating council and a physical framework plan to channel foreign assistance and investment into local infrastructure.

Selling places . . .

In addition to the incentives offered to investors and exporters, the Ramos administration also continued to market intensively the less tangible attractions of industrial location in the Philippines. These pronouncements are certainly toned down from the pitch of the martial law period two decades earlier, but some of the themes are the same. Several tropes stand out prominently in the government's 'place-marketing'.

Firstly, the 'strategic' geographical location of the country in the Pacific Asian or Pacific Rim region prompts an appeal to the boosterist tendencies of what Cumings (1993) calls 'Rim-speak'. Spatial metaphors abound, including the 'foothold' or 'gateway' to Asia, the 'heart' of the region, and a 'crossroads' for global flows and Eastern and Western culture.[5]

Secondly, the country is marketed as a conducive environment in which to do business with the support of a reliable and cooperative government. As Douglass (1993) points out, global investment decisions are as much about stable and supportive political environments as they are about specific locational incentives.

Philippine promotional literature attempts to assure potential investors of the government's commitment:

> The Philippines has been steadily and firmly putting in place all the elements needed to become an industrializing country: adoption of open-door policy; a spirited domestic and foreign investments drive; massive infrastructure development; government decentralization; tariff structure rationalization; a flexible exchange rate policy; vigorous export promotion and streamlining of export procedures; and import liberalization.
>
> (NEDA, 1995b)

Thirdly, promotional literature highlights the attractions of a Filipino workforce as a pool of low-cost, technically competent workers with English language proficiency. In this sense, like the model of 'Juan Kaunlaran' above, the Philippine government seeks to construct a Filipino identity that is attuned to the needs of the globalized economy. The components of this identity occasionally verge on orientalism, emphasizing the 'highly trainable' nature of the Filipino workforce.

Between 1992 and 1997, these policies allowed the Philippines to ride a wave of investment in Southeast Asia and economic indicators portrayed a booming economy as foreign investments flowed inwards and exports expanded. In 1994, in particular, substantial amounts of investment capital flowed into the country, and in 1994–7 GDP was once again growing at over 5 per cent. Estimates for 1998 suggest a slight economic contraction, but slow growth is predicted to restart in 1999.[6]

Clearly, numerous factors account for these trends since 1992, and the liberal Foreign Investments Act of 1991 and growing political stability are important. But, equally, longer term trends in *foreign* investment also show a close adherence to international economic cycles, with pronounced declines in the middle and late 1970s and again in the early 1990s. Similarly, growth in the mid-1990s reflects wider expansionary trends in the world economy and Southeast Asia in particular. This would suggest that current growth trends are susceptible to future recessionary conditions in the world or regional economy, as has been demonstrated in the 'contagion' effect of the Asian economic crisis. Furthermore, Philippine vulnerability to international economic trends, while it has always been high, has evidently increased substantially over the last decade. The ratio of trade to GDP provides a widely used indicator of openness of an economy to the world system. With a ratio of around 40 per cent, the Philippines now ranks as one of the most 'globalized' economies in the world.

. . . *Going places*

Beyond trade and investment policy, labour export also continued into the post-Marcos political economy and continues to be a mainstay of the national economy. Official government statistics recorded a total of 660,122 registered overseas contract workers deployed in 1996, but cumulative estimates put the actual number as high as 4.2 million (Ball, 1997; Gonzalez, 1998; NEDA, 1997). In the same year these workers remitted US$4.24 billion (NEDA, 1997).

It was only in 1995, after a massive public outcry following the execution of a Filipino maid in Singapore, that official attention was focused on the emotional and physical well-being of Filipinos working overseas. The case of Flor Contemplacion, accused of murdering another Filipino domestic worker and a Singaporean infant, starkly highlighted the harsh and sometimes cruel conditions in which overseas contract workers lived. But the depth of public emotion over the issue also indicated the very widespread identification with Contemplacion that existed throughout the Philippines. Taking one town in the province of Pangasinan as an example, Sheila Coronel (1998) points out that 3,500 people, out of a total population of 49,000 currently work abroad – 1 person in 14. Add to this the number who have worked abroad in the past, or hope to in the future, and almost every family has some immediate and personal interest in overseas work. Furthermore, this involvement goes far beyond economic considerations. The cultural impacts of overseas employment include: the reworking of gendered, sexual and class identities in families and communities in the Philippines; the effects of single parenthood on marriages and child rearing; a growing capacity in spoken English even among those with lower levels of formal education; and, less tangibly, a more cosmopolitan and expansive worldview among a broad segment of the population. While the latter point is difficult to substantiate, many would argue that overseas work in North America, Singapore, Hong Kong, and Europe has increased the demand for 'Western' brand-name products and in turn made policies of trade and investment liberalization an easier 'sell' for the Ramos administration.

The government's response to the Contemplacion affair was the hastily prepared Migrant Workers and Overseas Filipinos Act (Republic Act 8042) of June 1995. According to the stated intent, at least, of the law, the government would no longer explicitly rely upon overseas workers' remittances as a component of domestic economic policy (Gonzalez, 1998). With or without active government encouragement, however, overseas work continues to be an important part of the livelihood strategy for hundreds of thousands of Filipinos. The reason why this should be so while rapid domestic economic growth has proceeded in the mid-1990s is explained partly at least by the geographical and social distribution of that growth, which will form the focus of the next chapter.

Economic crisis and a populist president

The experience of regional economic crisis across Southeast and East Asia in 1997–8 exposed the vulnerabilities associated with a development strategy based on globalization. Considerable debate has followed over the extent to which domestic or 'global' causes have been behind the crisis. If a consensus is to emerge it seems likely that it will identify both as causes – global exposure accentuating and exploiting vulnerabilities, but vulnerabilities created by domestic macro-economic indicators and regulatory frameworks that do not conform to 'market expectations'. Thus, cronyism, fiscal deficits, short-term foreign loan exposure in the private sector, declining competitiveness, and exchange rate maladjustment may all represent proximal causes for the crisis, but it is the dependence of an economy on global capital, credit, commodities, technology and markets that makes these 'domestic' issues exploitable by global capital flows.[7]

As elsewhere in the region, the Philippines' domestic vulnerabilities were identified by currency traders and short term capital market players. These vulnerabilities included an increasing current account deficit, a growing dependence on foreign capital to finance growth due to the inadequacy of domestic savings, and a large real currency appreciation. These factors left the country susceptible to a currency devaluation, particular after the large falls in the value of other currencies, including competing export economies such as Thailand (Intal and Medalla, 1998). The peso exchange rate fell from just over 30 pesos to the US dollar in August 1997, to 45 pesos in early 1998, before settling at around 38 pesos in April 1998. The government's response to the crisis was to increase interest rates in an attempt to prop up the currency, but this lead to an economic slowdown as businesses faced increased credit costs. In turn, the government's already precarious fiscal situation worsened, with the national budget deficit doubling between the first quarter of 1997 and the same period in 1998 (Intal and Medalla, 1998).

The Philippines has not, however, experienced the widespread failure of financial institutions seen in Indonesia and Thailand, nor according to some analysts can the Asian Crisis actually be blamed for the a slowdown in economic growth. Intal and Medalla (1998) argue that while the crisis has affected the country's financial and fiscal sectors, its effect on national output and trade has been an aggravation of an existing decline, rather than the cause. They point out that growth was already declining in the first half of 1997 and lower rates were in fact precipitated largely by the lower yields in the agricultural sector caused by the El Nino climatic phenomenon. Meanwhile, export receipts were actually increasing during the year of 'crisis' between mid 1997 and 1998, assisted by the peso's devaluation.

How might the relative insulation of the Philippine economy from the 'contagion' that spread through the region in 1997/98 be explained? Firstly, the Philippine financial sector had already seen substantial reform in the 1990s, thus pre-empting many of the changes subsequently required under IMF duress in Thailand and Indonesia. These reforms included increased capitalization requirements, compliance with minimum asset ratios, limits on single borrowers, and stricter audit and reporting requirements (Intal and Llanto, 1998). Secondly, the country's macroeconomic management had been under IMF tutelage for the best part of twenty years. As a result, the national budget was in surplus, and its debt service burden (and proportion of short term to total debt) had declined significantly in the 1990s (Intal and Medalla, 1998). Thirdly, the process of rapid economic growth in the Philippines was at most five years old when the crisis struck and hence the forays of Philippine financial institutions into the international capital market were later (and less) than in other countries in the region (Intal and Medalla, 1998). Indeed, such was the relative health of the Philippine economy in 1997, some analysts claimed that the devaluation and consequent high interest rates were essentially a reflection of the 'herd instinct' and 'irrational pessimism' concerning East Asia that consumed the global financial markets from mid-1997 (Jurado, 1998).

One element of market confidence that the Philippines has been able to offer the global financial market in the last decade or so is political stability. While its regional neighbours suffered economic, social and political turmoil in 1997 and 1998, the Philippines conducted a largely free and fair set of elections in which the populist Joseph Estrada was elected to the Presidency. Estrada, who campaigned with the slogan 'Erap for the poor' (*Erap para sa mahirap*), had at first caused jitters among the business community which feared a rolling back of the Ramos administration's liberalization measures, but was eventually largely calmed by his choice of advisors.

The first few months of Estrada's administration have, however, created certain contradictions – a populist president with an anti-poverty agenda has been placed in a situation of declining revenues, fiscal cutbacks and lower growth. Estrada has suffered from the weight of massive expectations. Many believed that Estrada's 'pro-poor' platform would provide a growth-with-equity scenario in which the benefits of rapid development would be more equitably shared. Given economic circumstances, however, it seems unlikely that Estrada will be able to deliver on these expectations, and while the programme of deregulation, decentralization and liberalization seems unlikely to be unravelled, the roots of a globalized development path now looks a little less firmly embedded.

Constructing global imaginaries

What can be gleaned from this selective analysis of Philippine history? Firstly, it is evident that the basis of an economy and cultural consciousness oriented towards globalization can be traced, in part, to the legacies of colonial rule. In a variety of ways both Spanish and American regimes fostered an economy geared towards the export of primary commodities and the import of manufactured goods. Meanwhile, Filipinos were inundated with social and cultural hierarchies culminating not locally but in Madrid, Rome or Washington, and with images of the West as superior. All conducive cultural groundwork, as Constantino argues, for a development policy agenda dominated by the privileging of the global scale. At one level, then, the historical account in this chapter suggests some of the foundations of a contemporary political discourse predicated on globalization. But other themes also emerge that are germane to the key arguments outlined in chapter 1.

A significant legacy of colonialism was also an entrenched social structure, latterly legitimized by democratic processes and based on land ownership. Throughout the colonial period, precisely because they derived their wealth and prestige from it, this oligarchy of families carried a vested interest in the type of economy being established. Thus the globalization of the Philippines was not simply by colonial fiat, but was achieved with the collaboration of the local elite. Indeed a recurrent theme is the appropriation of outside power for domestic interests in the Philippines. In the second half of the twentieth century the social structure and economic interests of the domestic elite became more complex, but still development policy was defined through an articulation between outside influences and domestic political-economic interests. At a national level, then, the roots of globalization must be viewed as embedded in 'local' power relations rather than just the power of the outside actor.

Continuing the analysis of these themes into the contemporary Philippines we see the persistent influence of a societal structure embedded and reinforced by colonialism. But more significantly, the post-war and post-authoritarian political economy of the Philippines' place in the world has been shaped by interest groups within the Philippines jostling to make their interests the national interest. The articulation of the Philippine economy with the outside world has, however, taken on various forms over time, and has passed through phases of different levels of openness. What is now apparent, however, is that at no time in its history has the country been more open to international flows of capital, commodities, culture and people than it is at present. Furthermore, in recent years globalization has become not simply a description of change in the Philippine political economy, but also an argument in itself for particular economic policies. The idea of globalization has become more than just a way of understanding the world, it has become a way of changing it.

3
GLOBALIZATION AND THE PHILIPPINE SPACE ECONOMY
Patterns, processes and politics

Having established the broad nature of Philippine engagement with global flows in the previous chapter, we may now start to explore how this engagement has been inscribed upon the country's social and physical landscapes. This chapter will focus on the economic dimensions of globalization and particularly flows of foreign direct investment (FDI) in the export manufacturing sector. This is not to deny other important ways in which the country is integrated with the world economy, but FDI deserves special attention both because it has been a driving force behind recent Philippine development and because it forms a pivotal component of the government's globalized development strategy.

In the first section of this chapter a brief description of the regional structure of the Philippine economy will be provided, emphasizing the pattern of economic primacy in the national core region. The nature of direct investment flows that have been superimposed upon this spatial economic structure will then be examined, including their magnitude, sources, sectoral distribution and employment patterns. The third section of the chapter explores the spatial imprint of these foreign investments at three different scales. At the national scale, the concentration of investment in the national core region is highlighted. At the regional scale, within the national core, the key role of just a few provinces is demonstrated. Finally, within one such province, Cavite, an even finer spatial pattern is revealed – one that highlights a process of mega-urbanization or the emergence of an extended metropolitan region as a major spatial corollary of globalized development based on foreign investment flows. The fourth part of the chapter then attempts to provide an explanation for these spatial patterns, based on government policies, infrastructure provision, corporate strategies, and the local political economy of foreign investment.

The regional economic structure of the Philippines

As described in chapter 2, the colonial economy of the Philippines consisted of a series of regionally specialized economies producing primary commodities for export. The concentration of administrative and commercial activities, meanwhile, created a classic pattern of urban primacy centred on Manila. Despite efforts over the last three

Table 3.1 Regional share of GDP, 1981–96

Region	1981	1982	1983	1984	1985	1986	1987	1988	1989	1990	1991	1992	1993	1994	1995	1996
Central Luzon	9.1	9.2	9.2	9.2	9.4	9.1	8.8	8.5	8.1	8.5	8.6	8.7	8.7	8.4	8.4	8.4
NCR	29.1	29.5	29.8	28.7	28.7	29.6	30.6	31.4	32.3	32.3	32.8	32.4	32.6	32.1	32.5	32.8
Southern Tagalog	14.6	14.5	14.4	14.7	14.4	14.3	14.1	14.3	14.1	14.5	15.1	15.2	14.9	14.9	14.4	14.3
Northern Luzon	7.1	7.1	6.9	6.8	7.1	7.4	7.1	6.9	7.0	7.1	6.8	6.5	6.5	6.8	7.2	7.2
Bicol	3.3	3.3	3.3	3.3	3.4	3.3	3.0	3.0	3.0	3.0	2.9	2.9	3.0	3.0	2.9	2.9
Western Visayas	7.6	7.6	7.5	7.3	7.4	7.3	7.2	7.2	7.2	7.0	6.8	7.1	7.1	7.1	7.0	7.0
Central Visayas	6.4	6.4	6.4	6.4	6.3	6.3	6.3	6.5	6.7	6.6	6.6	6.6	6.5	6.5	6.4	6.4
Eastern Visayas	2.5	2.5	2.6	2.9	2.8	2.7	2.7	2.7	2.7	2.6	2.6	2.6	2.6	2.6	2.6	2.6
Mindanao	20.3	19.8	19.8	20.7	20.5	20.2	20.3	19.5	19.1	18.5	17.7	18.1	18.1	18.3	18.5	18.3
PHILS.	100.0	100.0	100.0	100.0	100.0	100.0	100.0	100.0	100.0	100.0	100.0	100.0	100.0	99.7	100.0	100.0
PHILS.('000 US$)	281596	317179	369078	524482	571884	608888	682765	799183	925444	1077237	1248011	1351559	1474457	1692932	1906328	2196595

Source: NEDA, 1997

Note: Regions in this diagram represent amalgamations of Philippine administrative regions and provinces. This takes account of changes in such regions over the past few decades, e.g. the creation of the Autonomous Region of Muslim Mindanao and the Cordillera Administrative Region. *Northern Luzon*: Regions 1 and 2 = Ilocos Norte, Ilocos Sur, La Union, Pangasinan, Batanes, Cagayan, Nueva Viscaya, Quirino, Kalinga-Apayao, Ifugao, Mountain Province, Benguet, Isabela, Abra. *Central Luzon*: Region 3 = Bataan, Bulacan, Nueva Ecija, Pampanga, Tarlac, Zambales. *Southern Tagalog*: Region 4 = Aurora, Batangas, Cavite, Laguna, Marinduque, Occidental Mindoro, Oriental Mindoro, Palawan, Quezon, Romblon, Rizal. *NCR* = National Capital Region (Metropolitan Manila). *Bicol*: Region 5 = Albay, Camarines Norte, Camarines Sur, Masbate. *Western Visayas*: Region 6 = Aklan, Antique, Capiz, Iloilo, Negros Occidental, Guimaras. *Central Visayas*: Region 7 = Bohol, Cebu, Negros Oriental, Siquijor. *Eastern Visayas*: Region 8 = Leyte, Southern Leyte, Eastern Samar, Northern Samar, Western Samar. *Mindanao*: Regions 9, 10, 11, 12, and Autonomous Region of Muslim Mindanao = Zamboanga del Norte, Zamboanga del Sur, Agusan del Norte, Bukidnon, Camiguin, Misamis Occidental, Misamis Oriental, Surigao del Norte, Davao del Norte, Davao Oriental, South Cotabato, Surigao del Sur, Davao City, Iligan, North Cotabato, Lanao del Norte, Sultan Kudarat, Tawi-tawi, Maguindanao decades to decentralize economic activity (described later in this chapter), the national core consisting of the National Capital Region (that is, Metropolitan Manila) and the adjacent Southern Tagalog and Central Luzon regions still accounts for more than half of the country's gross domestic product (see map 1).[1] This distribution is shown in table 3.1. In terms of Gross Regional Domestic Product, Manila has retained and even increased its share of national wealth over the 1980s and 1990s.

The core region's dominance becomes even more evident when manufacturing value added is isolated, as shown in figure 3.1. In manufacturing, the core region accounts for over two-thirds of the nation's activity, despite being home to only 38.5 per cent of the 1996 national population. While this level of dominance has remained consistent since 1981, a shift is also evident away from the Manila (the NCR) itself and into the adjacent regions of Southern Tagalog and Central Luzon. This trend has also been noted by Pernia and Israel (1994: 23), who provide the following commentary on regional trends between 1975 and1992:

> On the whole, the primacy of the NCR appears to be tapering off slowly in terms of industrial (manufacturing) activity besides population and employment. But the diffusion has been largely towards the surrounding regions, with some isolated regions making marginal gains the expense of others. The NCR continues to dominate in the production of services although its share in service sector employment has fallen somewhat, implying rising capital intensity in this sector. Further, Manila's primacy in terms of relative economic welfare seems to be on the wane. Some improvements are noticeable in nearby regions and even in the least developed ones, while the remote regions (Cagayan Valley [in Northern Luzon] and Mindanao) showed some deterioration in relative well-being.

Pernia and Israel also analyse inter-regional migration patterns and show that migration flows from the NCR into adjacent regions represent the most intense inter-regional movements in the country. Thus while labour mobility is high across the Philippines, in terms of both internal and international migration, it is the extended national core that provides the most attractive destination – a testament to its status as the country's most dynamic frontier of development. The resulting process of 'mega-urbanization' is a theme to which we will return later in exploring the geography of foreign direct investment.

Foreign direct investment flows

Chapter 2 described the various dimensions of the Philippines' integration into the global economy. Here, we will focus on perhaps the most significant, and certainly the most dynamic, aspect of this relationship: foreign direct investment (FDI) flows into the export sector of the economy. FDI has both flowed into, and shaped, the national space economy outlined above. Figure 3.2 shows the trend in capital inflows

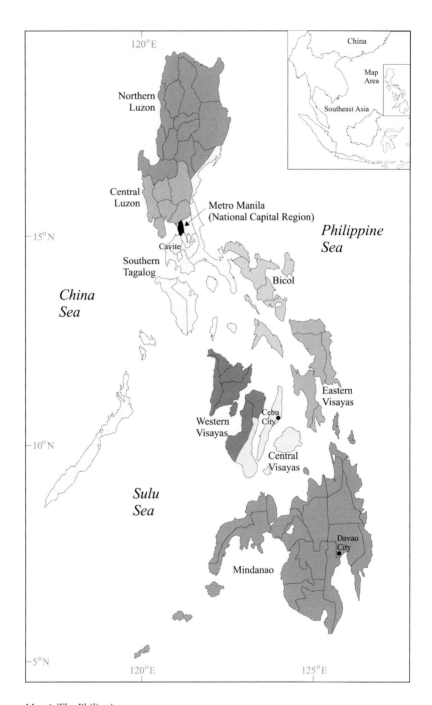

Map 1 The Philippines

GLOBALIZATION AND THE PHILIPPINE SPACE ECONOMY

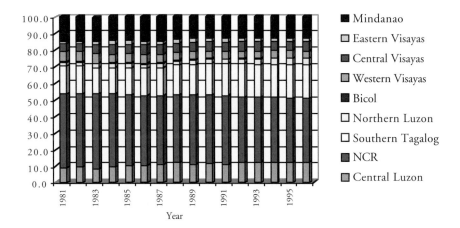

Figure 3.1 Regional share of manufacturing value added, 1981–96 (%)
Source: NSCB, 1997

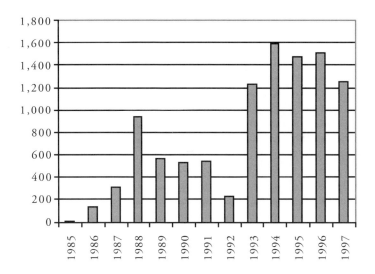

Figure 3.2 Foreign direct investment to the Philippines, 1985–97 (US$ millions)
Source: International Financial Statistics CD-ROM (IMF, 1998)

over the last ten years. The wave of investment that followed the downfall of Ferdinand Marcos in 1986 is clearly evident, but it was in the mid-1990s that massive flows of foreign investments poured in, encouraged by the liberalization policies and relative political stability fostered by the Ramos administration.

A comment is needed here on the issue of measuring FDI in the export sector of the Philippine economy. The direct investment data provided in figure 3.2 indicate actual capital flows into the Philippines from abroad as accounted by the Philippine Central Bank and reported to the IMF. Multiple sources and definitions of foreign investment exist for the Philippines; in this case, direct investment includes 'equity capital, reinvested earnings, and other capital associated with various intercompany transactions between affiliated enterprises' (International Monetary Fund, 1996: xviii). Excluded are flows of direct investment capital for 'exceptional' purposes such as debt-for-equity swaps, along with portfolio investments and loans.

Such figures are based on actual transactional flows of capital in a given year. They do not, however, correspond directly to the figures published by the various investment regulation agencies such as the Board of Investments (BOI) and the Philippine Economic Zones Authority (PEZA) which provide data on the cost of approved projects rather than actual capital inflows.[2] These agencies therefore exclude disinvestment and assume that all approved projects are actually implemented. Data from these sources has also been complicated in recent years by inconsistencies in definitions and reporting periods, often resulting in inflated investment figures. Only in 1997 did the National Statistical Coordination Board establish a Foreign Investment Information System to compile authoritative data on FDI. A further problem for present purposes is that data from all agencies combine investment from both local and foreign sources, and both export and domestic sectors.

Despite these problems, data from both BOI and PEZA will be used in this chapter because they provide a far more detailed picture of the sectoral and geographical breakdown of FDI flows than the Central Bank's figures. Where possible, however, PEZA data will be highlighted as its investments are, by definition, export-oriented and include a high proportion of foreign capital. In 1996, for example, 80.9 per cent of new investment in economic zones was from foreign sources, and by 1997 this had jumped to 96.3 per cent. For BOI registered projects, meanwhile, only 22 per cent of equity in 1996 was from outside the Philippines and only 3.7 per cent of such investment was in projects targeted at the export market (Medalla, 1998). This reflects the composition of BOI registered investment, which has seen a surge in public utilities and infrastructure projects in the 1990s, where foreign capital has moved in to take advantage of various frameworks established under the Ramos government's deregulation programme, such as build, operate, own (BOO) or build, operate, transfer (BOT) schemes.

Sectoral characteristics

The sectoral pattern of foreign investments in economic zones has been erratic over recent years, as table 3.2 shows. In general, however, the electrical machinery sector has been dominant, together with variable but occasionally significant contributions from other sectors including plastic products, transportation equipment, textiles, and garments (wearing apparel) (see plate 2).

Plate 2 A company manufacturing garments at the Cavite Export Processing Zone

Table 3.2 Investment in economic zones by sector, 1992–7

EPZ investment by sector (% distribution and million pesos)	1992	1993	1994	1995	1996	1997
Electrical machinery	21.9	13.5	66.7	73.9	61.4	74.1
Fabricated metal products	1.7	1.7	3.1	1.6	5.7	3.9
Industrial chemicals	4.3	1.4	0.0	1.8	0.5	0.5
Semiconductor manufacture & PCBs	0.0	0.0	0.0	0.0	0.0	6.7
Other equipment & instruments	8.5	13.1	0.7	2.2	0.3	0.3
Plastic products	15.0	4.5	9.3	0.9	1.8	2.1
Pottery, china & earthenware	0.0	0.0	0.0	0.0	0.04	1.2
Rubber products	4.2	6.8	1.8	0.5	0.4	0.7
Textiles	15.3	4.8	2.2	0.7	3.0	0.04
Transport equipment	1.7	14.2	8.3	8.7	11.9	3.1
Wearing apparel	13.0	9.2	4.3	0.3	2.2	0.5
Wood & wood products	0.0	0.0	0.1	2.6	1.3	0.2
Others	14.4	30.8	3.5	6.8	11.5	6.7
Absolute total (million pesos)	2365.3	2686.01	9590.2	44989.9	20457.2	52844.7

Source: Unpublished data from PEZA, 1998

Table 3.3 Sources of foreign investment approved by PEZA and BOI

	EPZA/PEZA		BOI	
	Million Pesos	Per cent	Million Pesos	Per cent
	(1992–7)		(1990–7)[a]	
American	12,850	11.0	48,880	22.5
ASEAN	4,059	3.5	25,540	11.8
Australian	534	0.5	1,350	0.6
European	7,466	6.4	52,820	24.3
HK	354	0.3	21,820	10.0
Japanese	69,613	59.9	27,310	12.6
Korean	13,581	11.7	4,200	1.9
Taiwanese	4,409	3.8	9,700	4.5
Other	3,427	2.9	25,600	11.8
Total	116,294	100.0	217,200	100.0

[a] Until October 1997
Source: Unpublished data from the Philippine Economic Zones Authority and the Board of Investments statistical website: http://www.sequel.net/~boimis/

Sources of Capital

Table 3.3 shows the level of foreign investment approved by both the Philippine Economic Zones Authority and the Board of Investments. The two sets of data show contrasting patterns. While export oriented investment registered with PEZA is dominated by Japanese, and to a lesser extent Korean and American, capital, projects registered with the BOI, which tend to be oriented more towards the domestic market are from a very different set of sources. American investment is still significant, but European, ASEAN and Hong Kong capital are far more important than in the export sector. Japanese investment is still a significant presence but far less than in export industries. An obvious inference is that Japanese and Korean investors are looking to the Philippines as an export base (a process well documented for other Southeast Asian countries in the aftermath of the Yen appreciation in the late 1980s – see Jomo, 1997), while American, European and ASEAN investors are more concerned with tapping into the growing domestic market.

Another more general point might also be deduced from these data. While overall FDI into the Philippines is impressively global and derived from all three 'triad' regions, export-oriented investment is dominated by Japanese firms. Such firms supplied almost 60 per cent of investment in economic zones in the early and mid-1990s. Thus, although it is reasonable to discuss recent Philippine development in terms of a 'globalizing' trend, the importance of a few key bilateral economic relationships should not be overlooked.

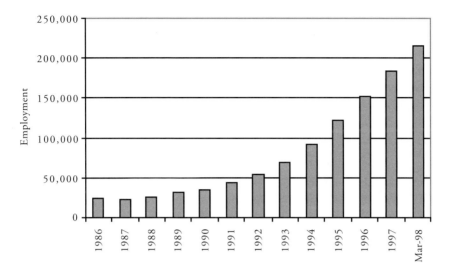

Figure 3.3 Employment in Philippine economic zones, 1986–98
Source: Unpublished PEZA data, 1998

Patterns of employment generation

The growth of employment in foreign-owned and export-oriented enterprises has been significant in recent years. Figure 3.3 illustrates this trend, which has been particularly pronounced since the Special Economic Zones Act of 1995 established an incentives structure for privately owned industrial estates.

Detailed data on the characteristics of the workforce in all PEZA zones or BOI projects are not available, but specific figures can be provided for the Cavite Export Processing Zone – the largest of PEZA's economic zones and the industrial hub that drives many of the processes of change that will be described in later chapters. These figures are shown in table 3.4. Garments and electronics companies dominate the zone, with stark gender differentiation within the workforce. Among foreign managers, men dominate in all sectors, while among Filipino management and staff the overall total is approximately equal, due mainly to female representation in the garment sector. The most dramatic gender differentiation is, however, at the level of production line employees. In all sectors except fabricated metal production, women constitute the vast majority of the workforce – 77 per cent of shopfloor employees in total.

At the CEPZ, workers do not live within the zone but are drawn from villages across nearby municipalities. Those who do not live with their families or relatives are accommodated in numerous boarding houses. In 1995, 20 per cent of workers at the Cavite EPZ lived in the town of Rosario itself – where the zone is located. A further 37.5 per cent lived in the adjacent municipalities of Noveleta, General Trias and

Table 3.4 Employment structure of CEPZ enterprises, April 1995

Sector	Management and Supervisory Staff				Employees		TOTAL
	Foreign		Filipino				
	Male	Female	Male	Female	Male	Female	
Wearing apparel	62	10	209	451	1,617	8,369	10,718
Athletic shoes and gloves	13	2	121	54	1,425	2,539	4,154
Hosiery	28	1	2	44	337	889	1,301
Plastic products	23	1	75	55	342	608	1,104
Fabricated metal	17	1	60	42	377	82	579
Electronics & electrical	189	5	1,283	888	2,798	12,040	17,203
Other	50	5	121	127	1,034	1,867	3,204
Total	382	25	1,871	1,661	7,930	26,394	38,264

Source: Unpublished data, Cavite Export Processing Zone, Industrial Relations Division, 1995

Tanza, while the remaining 42.5 per cent travelled from further afield (including 8.5 per cent from outside Cavite – almost certainly from Manila). The impact of the CEPZ and other smaller ones like it in Cavite is therefore widespread, with labour markets across the province affected by the significant employment opportunities provided.

The spatial patterns of FDI flows

Equipped with this sketch of the characteristics of FDI in the Philippines, we can now explore in more detail its geographical patterns. As will become clear, far from being a process that can be understood at a *national* level, integration into the global economy is highly uneven across the Philippine economic landscape. At a national scale there exists a pattern of concentration in the national core region. Within the core region, the provinces immediately adjacent to Metropolitan Manila have proven to be most dynamically transformed by global capital flows. Finally, even within these provinces, we see a pattern of development that suggests a mega-urbanization process emerging, with growth highly concentrated in relatively few municipalities. Each of these scales will be examined in turn.

National spatial patterns

The geography of new growth in the post-authoritarian Philippine economy can be indicated using data on employment generation in BOI registered investments.[3] In the latter part of the 1980s, the dominance of Manila was dramatic – over half of all new jobs created under BOI investments were located in the capital (see table 3.5). The national core region as a whole accounted for over 80 per cent of job creation. In the 1990s, the dominance of Manila has waned, but the core region as a whole

Table 3.5 Regional[a] employment generation by BOI projects, 1985–99

Region	% of 1990 Pop'n	1985–9	1985–9 % total	1990–4	1990–4 % total	1995–8	1995–8[b] % total
Central Luzon	11.6	29,984	7.2	50,777	11.8	40,952	12.7
Metro Manila	10.2	225,849	54.6	103,458	24.0	57,561	17.9
Southern Tagalog	13.1	81,054	19.6	136,883	31.8	93,388	29.0
Northern Luzon	13.6	2,484	0.6	15,946	3.7	12,197	3.8
Bicol	6.4	2,274	0.5	4,159	1.0	2,718	0.8
Western Visayas	8.8	12,879	3.1	9,691	2.2	6,629	2.1
Central Visayas	7.6	22,573	5.5	31,777	7.4	13,114	4.1
Eastern Visayas	5.0	533	0.1	3,414	0.8	873	0.3
Mindanao	23.6	32,225	7.8	49,412	11.5	33,503	10.4
Not indicated		4,036	1.0	10,149	2.4	2,042	0.6
Several locations		—	0.0	15,039	3.5	58,785	18.3
Total	100.0	413,891	100.0	317,624	100.0	321,762	100.0

[a] Regional definitions are provided in the note to table 3.1
[b] January–August only
Source: Calculated from unpublished data, Board of Investments, 1998; Population data calculated from NEDA, 1995c

remains pre-eminent with increasing shares of employment generation accruing to Central Luzon and Southern Tagalog. Gains in the rest of country were less dramatic, and the share of other regions has generally been far below their share of the national population.

The BOI figures used in table 3.5 include all investments, both domestic and foreign, that are registered under BOI incentives schemes for export producers or those operating in specified 'pioneer' sectors of the economy. As noted earlier, however, the majority of BOI registered investments are from domestic sources. Investments locating in Export Processing Zones and industrial estates administered by the Philippine Economic Zones Authority, on the other hand, are predominantly foreign investors, and all are export-oriented. The spatial pattern of economic zones therefore provides a clearer picture of how foreign direct investments are distributed within the Philippine space economy. Although PEZA documents recorded a total of eighty economic zones in January 1998, only twenty-three of these were actually active in terms of employment generation. These zones are shown in table 3.6.

The table indicates how dramatically export oriented foreign investment is concentrated in the national core region. Over three-quarters of all job opportunities in export firms registered with PEZA located in either Southern Tagalog or Central Luzon, with the former taking the lion's share. Thus, despite the existence of over eighty designated economic zones throughout the country, those that currently offer employment opportunities are heavily concentrated within the Southern Tagalog region.

Table 3.6 Distribution of employment in Philippine economic zones, 1998

Region/Economic zone	Employment	% of total
Central Luzon total	33983	15.7
Bataan EPZ[a]	27312	12.6
Angeles Industrial Park	354	0.2
Luisita Industrial Park	3613	1.7
Victoria Wave Special Ecozone	2033	0.9
Subic Special Ecozone	671	0.3
Metro Manila (NCR) total	0	0.0
Southern Tagalog total	131295	60.4
Cavite EPZ[a]	54141	25.0
Carmelray IP	13048	6.0
Cocochem IP	25	0.0
First Cavite IE	6183	2.9
First Philippine IP	153	0.1
Gateway IP	8677	4.0
Laguna International IP	4244	2.0
Laguna Technopark	23397	10.8
Light Industry & Science Park I	18366	8.5
Light Industry & Science Park II	1336	0.6
Lima Technology Centre	1245	0.6
Toyota Industrial Complex	480	0.2
Northern Luzon total	5118	2.4
Baguio City EPZ[a]	5118	2.4
Bicol total	0	0.0
Western Visayas total	0	0.0
Central Visayas total	38381	17.7
Mactan EPZ[a]	36252	16.7
Mactan Ecozone II	1879	0.9
New Cebu Township	250	0.1
Eastern Visayas total	7651	3.5
Leyte Industrial Development Estate	7651	3.5
Mindanao total	110	0.1
First Oriental IP	110	0.1
Total	216538	100.0

Note: Those marked with '[a]' are government owned. All others are private, but registered with and regulated by the Philippine Economic Zones Authority
Source: PEZA, 1998, unpublished data

Regional spatial patterns

Within the national concentration of economic growth, there also exists a more subtle spatial pattern. As regards the employment generated by BOI-registered investment in Southern Tagalog, for example, table 3.7 shows that the vast majority (over 80 per cent in 1995–8) of jobs were concentrated in just two provinces – Cavite and Laguna (see map 2). The importance of these two provinces in the national space economy is immense. If we calculate their employment generation in BOI projects as a percentage of the national total (excluding those investments with

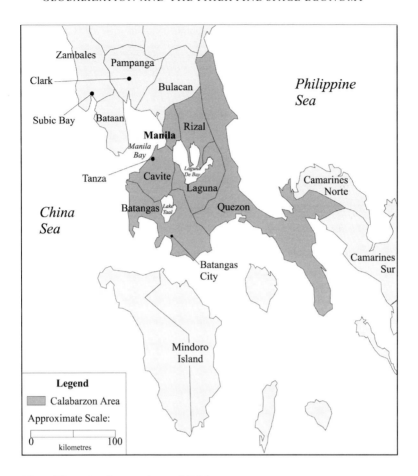

Map 2 The national core region and Calabarzon area

no fixed location) they account for 10.9 per cent in 1985–9, 30.6 per cent in 1990–4 and 29.3 per cent in 1995–8. Thus, in the mid-1990s, two provinces, with a combined share of the national population of just 3.2 per cent in 1990, were accounting for almost one-third of all employment generation in investments registered with the BOI.

A similar pattern of concentration is evident in the location of employment in PEZA registered economic zones. Without exception, all of the zones and industrial estates listed above in table 3.6 under Southern Tagalog are in Cavite and Laguna. Thus, just over 60 per cent of all PEZA registered employment generation is located in these two provinces. Clearly, together they represent the most dynamic zone of industrial development in the country and the provinces most closely linked with flows of foreign direct investment.

Table 3.7 Employment generation by BOI-registered investments, 1985–98 in southern Tagalog provinces

Province	1985–9	%	1990–4	%	1995–8	%
Batangas	4,214	5.2	12,266	9.0	8,075	8.6
Cavite	13,882	17.1	48,301	35.3	55,984	59.9
Laguna	30,751	37.9	41,088	30.0	20,448	21.9
Oriental Mindoro	0	0.0	803	0.6	68	0.1
Quezon	742	0.9	3,474	2.5	975	1.0
Rizal	25,435	31.4	25,149	18.4	6,279	6.7
Palawan	627	0.8	4,820	3.5	1,254	1.3
Aurora (sub-province)	0	0.0	200	0.1	0	0.0
Romblon	3,500	4.3	221	0.2	0	0.0
Occidental Mindoro	421	0.5	484	0.4	53	0.1
Marinduque	1,482	1.8	77	0.1	0	0.0
Southern Tagalog	81,054	100.0	136,883	100.0	93,388	100.0

Source: Unpublished data, Board of Investment, 1998

Provincial spatial patterns

Having established the pre-eminence of Cavite and Laguna within the Philippine space economy it is possible to identify even finer geographical patterns of industrial development. Taking Cavite in particular, table 3.8 provides data on the province's industrial sector shown according to municipality (see map 3).

Over 96 per cent of all industrial employment in the province in 1995 was located in the northern lowland towns closest to Manila. In particular, the towns of Carmona, Dasmarinas, Imus, General Trias and Rosario support the bulk of the province's industrial sector. All are located on the agricultural fringe of Manila, beyond its suburban sprawl into Bacoor and Noveleta, and all are within the prime agricultural areas of the province's lowlands. While various plans have sought to direct industrial development into satellite centres, in a 'leap-frogging' pattern, the reality is that industrialization is being concentrated in an extended metropolitan region around Manila. The capital's hinterland is thus witnessing a proliferation of industrial and residential developments juxtaposed with agricultural land.

The pattern of development described here is far from unique to the Philippines. The resulting landscape of intensively mixed agricultural and urban/industrial activities around a major city has received considerable attention in recent years as analysts have attempted to come to terms with an Asian process of urbanization quite distinct from the Western experience. These extended metropolitan regions have been dubbed 'desakota' by McGee (1989), combining the Bahasa Indonesia words for village (*desa*) and town (*kota*) to indicate both the integration of the two in both function and landscape, and to highlight the need to use local terms instead of Western categories in understanding Asian urbanization (McGee, 1991). According to McGee, such 'desakota' regions represent:

Table 3.8 Characteristics of Cavite's industrial sector, June 1995

Municipality / Industrial Estate	Number of companies	Capitalization (million Pesos)	Employment
Alfonso	4	246	277
Bacoor	9	109	885
Carmona	78	5,504	10,613
People's Technology Complex	*45*	*3,596*	*6,560*
Granville Industrial Complex	*10*	*168*	*1,383*
Mount View Industrial Complex	*8*	*422*	*988*
Southcoast Industrial Estate	*4*	*7*	*225*
Outside industrial estates	*11*	*1,311*	*1,457*
Dasmarinas	41	4,134	6,431
First City Land Heavy Industrial Centre	*2*	*10*	*39*
DBB-NHA Industrial Estate	*4*	*71*	*2,448*
First Cavite Industrial Estate & SEPZ	*21*	*1,863*	*2,382*
Outside Industrial Estates	*14*	*2,190*	*1,562*
General Mariano Alvarez	6	272	1,109
GMA-NHA Industrial Estate	*5*	*110*	*1,004*
Outside Industrial Estates	*1*	*162*	*105*
General Trias	22	5,022	2,512
New Cavite Industrial City	*10*	*171*	*610*
Gateway Business Park and SEPZ	*4*	*2,787*	*987*
Outside Industrial Estates	*8*	*2,064*	*915*
Imus	16	2,658	8,916
Indang	4	201	—
Naic	9	70	344
Noveleta	2	3	227
Rosario	143	4,667	39,347
CEPZ	*140*	*4,591*	*38,264*
Outside CEPZ	*3*	*76*	*1,083*
Silang	15	755	2,279
Tanza	4	1,460	860
Cavite City	4	11	160
Tagaytay City	5	20	123
Trece Martirez City	14	104	806
Total	376	25,236	74,889

Source: Unpublished data, Province of Cavite, Planning Department, 1995

Distinctive areas of agricultural and non-agricultural activity emerging adjacent to and between urban cores, which are a direct response to pre-existing conditions, time–space collapse, economic change, technological developments, and labor force change occurring in a different manner and mix from the operation of these factors in the Western industrialized countries in the nineteenth and early twentieth centuries.

(McGee, 1991: 4)

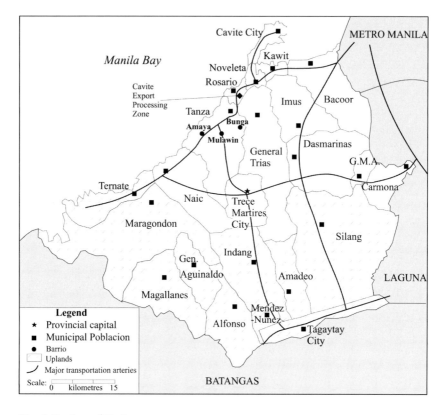

Map 3 Province of Cavite

Thus the 'desakota' pattern of urbanization represents a distinct spatial configuration involving not simply either densification of the urban core nor progressive outward suburban growth, but rather a region-based pattern of urbanization to form an extended metropolitan zone. These are zones of changing land use and economic structure, experiencing major transformations in employment patterns, family and social life, local politics and power relations, and the physical environment.

The theoretical literature on mega-urbanization (Ginsburg, Koppel and McGee, 1991; and McGee and Robinson, 1995) has attempted to explain these 'urban' forms through a variety of situational and historical factors that have made the Asian developmental experience distinct from the West.[4] A frequently cited factor is the location of Southeast Asia's large cities within densely populated rural areas based on labour-intensive wet rice cultivation. This juxtaposition means that a large potential workforce (and market) for the industrial and service sectors exists beyond formal city limits, thus the need for urban agglomeration is circumvented in such areas. The second factor, which largely results from high population densities, is the intense 'transactional environment' in agricultural hinterlands facilitated by relatively inexpensive and varied modes of transportation such as motorized tricycles, becak, trishaws, jeepneys, buses, and motorized canal boats. These allow the easy circulation of commodities, capital, and particularly people, and thus once again negate the requirement for urban agglomeration.

In addition, what makes recent Southeast Asian urbanization distinct is the driving force provided by the surge in export-oriented and foreign-investment-driven industrialization in national core regions described earlier. Contemporary mega-urbanization patterns might, therefore, be seen as the spatial corollary of a globalized development strategy. The next section of this chapter will attempt to explore the connections between globalized development and the spatial pattern described.

Explaining the geography of a globalizing Philippine economy

As noted, the globalizing economies of Southeast Asia have all experienced a spatial pattern of growth similar to that of the Philippines. Jakarta, for example, represents only around 3.6 per cent of the Indonesian population but received 17.2 per cent of the cumulative value of foreign investment between 1967–91 (Soegijoko 1996; see also Forbes, 1986, and Forbes and Thrift, 1987, on Indonesia). If surrounding areas in West Java are included the share rises to 44.8 per cent of the national total. This translates into more than half of the jobs created by foreign investment in the same period. In Thailand, 67.8 per cent of projects approved by the Board of Investments between 1979–90 were in the Greater Bangkok Metropolitan Region (Medhi, 1996). Vietnam's transitional economy, subject to intense foreign capital flows more recently than the market economies of the region, displays a similar pattern. Nestor (1997) shows that for the period 1988–93, 45.9 per cent of foreign joint venture investment was in Ho Chi Minh City, with a further 21.1 per cent in Hanoi.

The Philippines' pattern of spatial development is not, therefore, unique. There is, however, a distinctive set of local geographical and institutional explanations for the pattern revealed above. In this section I will outline four contributing factors that seem to lie behind the differential imprint of globalized development on the Philippine landscape: government policy; infrastructure provision; corporate requirements; and local politics.

Government policy

An obvious starting pointing for explanations of the geography of Philippine development is with government regional policy. It has, however, been the ineffectiveness and constant undermining of spatial planning, rather than its impact, that emerge as its most significant role.

Until the 1960s, regional policy was not a significant issue in Philippine development planning (Pernia, 1988). Agricultural processing and export had been dispersed across the various regional economies specializing in different crops, and import-substituting manufacturing was concentrated around the prime market of Manila (McGee, 1967; Pernia and Paderanga, 1983). Only in the 1970s did government policy start to contain an explicitly spatial component, with benefits for regional dispersal being added to the incentives for investment and export production. By locating in certain BOI-designated areas, companies could receive a doubled tax credit for direct labour costs and a tax deduction for the cost and maintenance of infrastructure work undertaken by the firm (Reyes and Paderanga, 1983). Other regional dispersal incentives were effectively provided through the export processing zones, three of which were located well beyond the capital (Bataan, Mactan and Baguio City). Despite these incentives, investment and productive activity in the 1970s continued to be concentrated in the Manila region. Between 1970 and 1977, 73 per cent of firms registered under the Export Incentives Act (RA 6135) were located in Metro Manila, with a further 12 per cent in the two adjacent regions of Central Luzon and Southern Tagalog (Reyes and Paderanga, 1983).

In 1973 a further and more explicit regional policy initiative aimed at decongesting Metro Manila was introduced (Reyes and Paderanga, 1983). New manufacturing plants would need a locational clearance from the Ministry of Human Settlements (headed by Imelda Marcos), and there would be a ban on locating within a 50 km radius of Manila (of which Mrs Marcos was the governor). The result of this ban, however, was the location of many factories on the edges of the exclusion zone where they could still take advantage of the transportation infrastructure and market in the capital. Thus in the period 1974–8, 48 per cent of locational clearances were in the Central Luzon and Southern Tagalog regions, and despite the regulations, a further 28 per cent were exemptions granted to locate within the National Capital Region (Reyes and Paderanga, 1983).

Another explicitly spatial development policy in the 1970s and early 1980s was the establishment of Integrated Area Development Projects. These projects, such as the Bicol River Basin Development Project, were intended to coordinate infrastructure programmes in various regions and direct them towards the requirements of industrial development. Despite these programmes, Pernia and Paderanga (1983) argue that incentives for regional dispersal were outweighed by the implicit concentrating tendencies in export promotion policies. The need to be near government offices, major banks and transportation facilities meant that most firms located within the core region.

Further efforts at administrative and industrial decentralization were undertaken by the Aquino administration. Perhaps the most significant has been the programme of administrative devolution initiated by the Local Government Code of 1991 (RA 7160). Under the code, local government units (municipalities and cities) were granted greatly increased powers in land use classification and planning, thereby allowing them to control a key factor of production. Responsibility for approving development plans for residential subdivisions, commercial premises or industrial estates no longer rests with the National Housing and Land Use Regulatory Board, but with local government officials. The result is that municipalities are free to establish industrial zones within their jurisdictions and to try to attract industrial investment. But, as all municipalities have this capability, it is those with adequate infrastructure and with the closest links to transportation and communications facilities that can benefit. The result has been that municipalities closest to Manila have been most successful in this respect.

Another form of regional dispersal policy has been pursued through the Export Processing Zones, and, more recently, through the growing number of private industrial estates listed in table 3.6. These estates can provide incentives such as tax exemptions or deductions, waivers on local taxes, simplified export procedures, and existing infrastructure (Lamberte et al., 1993: 44). As noted earlier, the national impact of such estates on regional dispersal is limited, however, since most are located within Region IV in the provinces of Cavite and Laguna.

Other growth hubs include the redeveloped US military bases at Subic (Olongapo City) and Clark (Angeles City). The former naval base at Subic Bay has been converted into a free port and industrial estate. The area has positioned itself as a hub for regional headquarters in Southeast Asia and by May 1997, the Subic Bay Metropolitan Authority had approved investment projects worth 52.45 billion pesos (then approximately US$2.0 billion) since its inception in 1992. By the same point, Clark had brought in 29.97 billion pesos since 1993.[5]

The most prodigious of the Philippines' developmental regions, however, has been the area around Manila itself. As noted earlier, Manila has developed in a pattern also exhibited by other Asian cities, including expansion into a zone of intensively mixed agricultural, residential and industrial uses. This zone has become known as Calabarzon – an acronym that groups the provinces of Cavite, Laguna, Batangas, Rizal and Quezon lying to the East and South of the capital (see map 2). The project was initiated in 1990 by the Department of Trade and Industry as a coordinated effort to transform agrarian provincial economies into urban-industrial ones through export-oriented industrialization (JICA, 1991). The Philippine government sought technical assistance from Japan and in 1991, the Japan International Cooperation Agency produced the 'Master Plan Study on the Project Calabarzon', a thirteen-volume report on all aspects of the area's future development. The plan provides a framework for development until 2010 in three principal categories: the 'socio-economy' (intensification of agriculture, industrialization, tourism, and human capital development); infrastructure (highway construction, port development, housing, water and power supply, and industrial estates); and spatial development. The funding for these developments is envisaged as coming from the Philippine Assistance Plan, particularly Japan's Overseas Economic Cooperation Fund, and from private investors. The projected public expenditure for various projects was estimated at US$615 million in 1991–5 and then US$713 million in 1996–2000.

The spatial development framework favoured by the Calabarzon masterplan involves moderate expansion of Manila's suburban fringe coupled with 'leap-frog' development in secondary industrial centres such as the port city of Batangas (JICA, 1991; Laquian, 1996). While the area has been successful in terms of attracting investment, as a spatial planning exercise its vision has been a decided failure. As the provincial data provided earlier showed, development has in fact been concentrated in the provinces and towns closest to Metro Manila. Part of the reason for this appears to lie in the institutional structure of the plan. As an entity, Calabarzon exists in the form of a Coordinating Council and a small secretariat, but control over development planning effectively rests with the respective provincial governors who form the council. The result has been that provincial politics and development agendas (described below) have been more influential than the regional framework recommended by the JICA planners.

In various ways, then, the regional planning efforts of the Philippine government appear to have had little impact on the country's spatial development. While some projects, such as the siting of an export processing zone in Bataan or the designation of industrial estates with location incentives in Mindanao, have attracted investment that might not otherwise have located in outlying regions, the overwhelming thrust of development has been in the national core region.

Infrastructure

While regional policy engages in an explicit spatial planning strategy, an implicit and more influential component of public policy relating to spatial development is infrastructure provision. Spatial inequalities in the provision of physical needs such roads, telecommunications, power, air and seaports reflect an explicit attempt to be responsive to the needs of foreign capital, resulting in public investment disproportionately concentrated in the core region. As table 3.9 shows, the Southern Tagalog region featured prominently in public infrastructure investment in the period 1989–92, and future plans indicate that its share of regionally designated programmes will remain dominant.

Such a concentration of public infrastructure investment is found, for example, in transportation projects. The Medium Term Philippine Development Plan 1993–8 (NEDA, 1995a) lists eleven major transportation projects under the BOT scheme with an estimated total cost of 82.1 billion pesos (approximately

Table 3.9 Regional distribution of public investment, 1989–92 (%)

Region	Share of region-specific public investment, 1989–92	Share of total public investment, 1989–92	Future projection
Northern Luzon	4.9	2.01	4.71
Central Luzon	15.6	6.34	7.90
Metropolitan Manila	22.2	9.04	4.83
Southern Tagalog	18.4	7.48	16.68
Bicol	8.0	3.26	2.72
Western Visayas	0.9	0.36	2.43
Central Visayas	5.1	2.08	0.84
Eastern Visayas	9.8	4.00	9.94
Mindanao	15.0	6.09	6.62
Inter-regional	—	10.66	6.82
Nationwide	—	48.67	36.53
Total	100.00	100.00	100.00

Source: NEDA, cited by Lamberte *et al.* (1993)

US$4.1 billion in 1995). All eleven are in Metro Manila or its adjacent regions and include the capital's Light Rail Transit System, the Manila Circumferential Road and the Manila–Cavite Expressway. Similarly, in 1996, 76 per cent of telephone lines were in the national core region, and 58 per cent were in Metro Manila alone (NEDA, 1997).

As Pernia and Israel point out, 'public infrastructure policy, in practice, tends to favour prosperous areas over lagging regions, and industry over agriculture, thus resulting in ever widening interregional gaps' (1994: 24). What has emerged, then, is a policy of what might be termed 'spatial trickle down', focused on creating a core region attractive to foreign investors, pending the eventual presumed expansion of economic growth to peripheral hinterlands.

Corporate locational requirements

Parallel to the spatial unevenness of infrastructure provision are the locational requirements of foreign investors entering the Philippines. Few extensive firm-level studies of investors have been conducted, but available information suggests a predictable list of requirements. In a study of 100 firms, Herrin and Pernia (1987) found that the common needs of firms in the Philippines included proximity to customers, easy road access, reliable electricity supply, adequate telecommunications facilities, suitable land availability, and space for expansion. According to Herrin and Pernia, regional labour cost variation is generally not a critical locational consideration for choice of location within the Philippines. Furthermore, while government guidance or encouragement can exert some influence, it is far less decisive that the other factors listed (see also Lamberte *et al.*, 1993: 33). There is, however, a distinction to be made between export oriented and domestic-market firms and between local and foreign capital. Of primary interest here is export-oriented foreign capital, which Herrin and Pernia suggest tends to focus on the availability of a labour force with appropriate skill levels, access to air and sea ports, access to plant sites, and the cost of land. Pernia and Paderanga (1983) also suggest the importance of locations close to government offices, major banks, and professional services. Since all of these facilities (as well as domestic market demand) are overwhelmingly concentrated in Manila and surrounding provinces, the core region retains its attractiveness.[6] Here, the dense populations and intense 'transactional environments' noted in the literature on extended metropolitan regions are clearly important factors. In addition, while the national core dominates in the satisfaction of formal business requirements, it is also likely that political instability in Mindanao and periodic kidnappings of foreign nationals in peripheral regions provide strong personal reasons why foreign investors have attempted to avoid such areas.

Local politics

In reality, corporate locational needs and government responses provide only a partial picture of the processes involved in the geography of Philippine industrial development.

Even when the issue is growth management rather than growth generation, as in the Calabarzon regional framework, the data already presented on industrial expansion suggest that public regulation has been largely ineffectual. It would seem that plans have been consistently contravened and subsumed under an overarching imperative to attract foreign investment *wherever* possible. The process through which this contravention occurs lies, at least in part, in the highly politicized nature of local economic planning. Among other provisions, the Local Government Code of 1991 allows municipalities and provinces to take a highly proactive role in land use planning, infrastructure development and investment promotion, meaning that each attempts to maximize its economic base (and the enrichment of local political leaders), often in disregard for broader planning frameworks. In Cavite, for example, a province that according to the Calabarzon Master Plan should be preserving its agricultural areas, large tracts of irrigated rice land have been converted to housing or industrial uses due largely to the context of industrialization in a complex local political economy. In this section, and then in more detail in subsequent chapters, I will focus on Cavite and consider the ways in which national development strategies (implicit or explicit) are mediated at smaller scales.

To understand the role of local political processes in economic development it is important to consider the historical context of power relations in a locality such as Cavite. Three important features of Cavite's political culture and economy can be identified as emerging from Spanish colonialism. Firstly, a rigid class structure (albeit with some potential for mobility) based on control over land was formed. It consisted of: wealthier tenants (*inquilinos*), who rented directly from the owners of the land (usually Spanish religious orders which acquired large tracts in the northern lowlands of the province); impoverished subtenants (*kasamas*); and, landless labourers (*jornaleros*) whose position was economically and socially vulnerable and had little legal standing (Roth, 1982). Secondly, these social relationships were defined through the tenure system, thus making land ownership the key to broader economic and political power. Thirdly, the social inequities formed by this system fostered a culture of banditry and political violence for which Cavite remains notorious in Filipino popular culture (Medina, 1994).

In a series of essays, John Sidel (1994, 1997, 1998; see also Kelly, 1997, 1998) provides a compelling picture of how Cavite's historical evolution has shaped its contemporary political economy. The province has always benefited from relative wealth, derived from its proximity to Manila, a fertile rice growing belt in the Northern lowlands, a coffee and fruit based sector in the southern uplands, and fishing industries and smuggling in coastal villages. In these activities two distinct characteristics can be identified, also present in the rest of the country but especially prominent in Cavite. The first is the intersection of legal and illegal forms of both political and economic power; the second is the overlap between processes of public regulation, private accumulation and such political power. Numerous examples exist of the machinery of the state being used to regulate economic activities for personal gain. These economic activities may include both legitimate and illicit opportunities.

The integration of politics, economics and criminality in Cavite is illustrated by activities in the municipality of Tanza in the 1950s and 1960s. In a period when protectionist policies imposed tariffs on imported goods, Tanza's coastal location made it a prime site for smuggling goods (particularly cigarettes and firearms) from the north coast of the island of Borneo (shared by Malaysia and Indonesia). Capipisa, one of the municipality's coastal barrios, earned a reputation for its ostentatious residences and well-paved streets. These operations grew to a considerable scale, with the distribution of smuggled goods extending into Manila, under the acquiescence of government officials at all levels. Indeed, according to Sidel (1995: 363), protection from leading politicians – notably Governor Delfin Montano of Cavite, his father Senator Justiniano Montano, and their allies – was essential in diverting the attention of law enforcement agencies. Military units 'not only turned a blind eye to the syndicate's smuggling operations but in fact provided the coercive resources of the state for the policing and enforcement of the syndicate's monopoly against potential competitors' (Sidel, 1995: 364). This system of protection broke down with the election of Ferdinand Marcos in 1965 who harboured considerable enmity towards the Montanos. But rather than closing down the operation, Marcos simply inserted himself as its patron and ensured the victory of the syndicate's leader, Lino Bocalan, in the election for provincial governor against Delfin Montano in 1971 (Sidel, 1995: 368).

In Cavite, then, the machinery of local government has become a powerful tool for personal wealth accumulation. Monopoly control over such machinery by elected local politicians creates what Sidel (1995) refers to as 'bossism'. Many mayors, for example, have achieved considerable wealth through the judicious application of their powers of office. This extends well beyond the local political control of law enforcement and includes the abuse of regulatory, employment, revenue dispersal and contract-granting powers. The result is that 'municipal mayors, through their institutionalized control over the local coercive and extractive apparatuses of the state, exercise considerable regulatory powers over the – legal and illegal – economies of their respective municipalities' (Sidel, 1995: 230).

The application of these powers has reached even higher stakes with the rapid southward suburbanization of Manila and the development of export-oriented industrial developments in the province over the last decade. In particular, control over land use and zoning ordinances has become the prime source of regulatory leverage over the formal economy open to municipal mayors in Cavite. As pressure grows for conversion of agricultural land into residential subdivisions, industrial estates, commercial developments and leisure facilities, mayors are also able to exert influence through the issuance of building permits, the use of government lands, the allocation of public works, and the implementation of agrarian reform (Sidel, 1995: 246). Agrarian reform is especially important because designation for redistribution under the programme would prevent the conversion of farmland to other uses. 'With such mechanisms of land-use regulation at their disposal, Cavite's municipal mayors have evolved into the province's leading real estate agents and brokers' (Sidel, 1995: 247).

Urban and industrial development, then, are processes in which local politicians have a considerable vested interest. It is in this context, in which leaders at the municipal and provincial level have access to monopoly powers of regulation and coercion, that the process of foreign direct investment must be viewed. The provincial level of government appears to have played an especially decisive role in attracting foreign investment to Cavite and one man in particular has been widely identified with the aggressive promotion of Cavite as destination for investment: Governor Juanito Remulla (1979–86 and 1988–95).

A native of Imus, Remulla attained the governorship of the province after President Marcos (his fraternity brother at the University of the Philippines) persuaded him to renege on his allegiance to the locally powerful Montano family.[7] Remulla enjoyed close ties with Marcos and facilitated business ventures in Cavite for several prominent 'cronies', but his position was also strengthened by strong links with the powerful Puyat family in Manila, whose interests extended from banking to manufacturing and, of course, politics. In addition to these powerful connections beyond the province, Remulla retained a formidable political machine within his jurisdiction. Many of Cavite's mayors, councillors, provincial board members and congressmen have owed their electoral successes at least in part to their status as Remulla's protégés. In the local elections of 1980, 1988, 1992, and even in 1995 when Remulla lost his office, over three-quarters of the province's elected mayors have been on the Remulla 'slate' (Sidel, 1995). These political ties have in turn translated into mutually beneficial economic opportunities in the public, private and 'informal' sectors.[8] As Alfred McCoy points out with reference to the Philippines in general, 'political families at both the provincial and national levels thus share common involvement in rent-seeking politics, a process of turning political capital into commercial opportunity' (McCoy, 1994a: 22).

Remulla's strategy for attracting investment took three specific forms. Firstly, the provincial government engaged in independent and aggressive place-marketing campaigns to attract foreign capital. These included promotional missions overseas, the hosting of visiting trade delegations, and the publication of brochures in English, Japanese and Chinese. Secondly, local government units, including both provincial and municipal administrations, have become heavily involved in the process of land conversion to provide for the needs of industrial developers. Thus local politicians have been able to assure investors that their needs for industrial sites will be met. Thirdly, local politicians have sought to ensure that the labour process within their jurisdiction has been conducive to the needs of foreign investors. This has meant securing a workforce of sufficient size (by allowing extensive developments of low cost housing) and low cost (by resisting moves towards unionization).

In these three ways, then, local political actors can play a prominent role in constructing an environment attractive to foreign capital. While it is arguable that opportunistic local power brokers are simply 'cashing in' on the rapid urban and industrial development that comes their way, it must also be recognized that local political economic structures provide a significant input to the development process. Land and labour form two of the key factors of production on which direct investment is predicated; they are also factors that are essentially 'local' in their regulation. The ability of local political actors to secure these factors in appropriate quantities and qualities, often in the face of local resistance, must be seen as part of any explanation of geographical patterns of investment. In several places in the Philippines, the local political economy has been decisive in this way – Subic Bay under the guidance of Richard Gordon and Cebu under the Osmena family are two clear examples (Chan, 1999; Sidel, 1998). It has been in Cavite, however, that in the words of one of the province's investment promotions brochures, 'the sociopolitical setup . . . [has] produced the consensus for a single development-oriented team' (Province of Cavite, n.d.: 1). There, the powerful imbrication of politics and economics means that an understanding of local systems of power relations is essential to an account of recent trends in the province's economic development and its articulation with the global economy. It is the political economy of land and labour in Cavite to which we will turn in the next two chapters.

Conclusion

A striking feature of recent growth in the Philippines has been its spatial distribution, with a high concentration in the core region, and particularly in the vicinity of Manila. In part, this obviously reflects the locational preferences of foreign investors (Pernia, 1988), but two other factors have also been noted.

Firstly, a priority of the government, based on its reading of globalization, has been to attract as much investment as possible, with goals of regional equity effectively marginalized. While the government seeks to position the country internationally to attract investment, the spatial corollary at a national level has been a concentration of economic activity in the core area. The government has thus allowed industrial development to locate in the core region that is preferred, for various reasons, by international investors. This preference has been further reinforced by public infrastructure investment that is also heavily concentrated in the national core. Thus despite decentralization policies, it has been the provinces adjacent to Manila that have attracted the bulk of recent investment flows. In terms of spatial equity, then, the consequences of a globalized development strategy are highly political. This supports one of the two key arguments, concerning the political nature of globalization discourse, presented in chapter 1.

The other key argument that emerges is the importance of local social relations in shaping experiences of globalization. The account of policies and economic performance provided in chapter 2 indicated that this has been true at a national level. The corrupt, personalistic and kleptocratic tendencies of the Marcos regime make that point clearly – even while the Philippines was pursuing policies that copied its successful neighbours, its globalized development strategy was undermined by the system of economic power relations that held sway during the 1970s and early 1980s. But as the final section of this chapter suggests, political power at other scales has also been influential in determining the economic geography of a globalizing economy. A key component of the national government's programme of decentralization has been the devolution of various powers and responsibilities to other administrative scales, for example through the Local Government Code of 1991. This, combined with the networks of political power that already existed at municipal and provincial levels, has allowed these scales to find an active role in shaping the localized process of globalization. The ways in which these political economic relations have been played out in the social and physical landscapes of Cavite form the subject of the next two chapters.

4

THE GLOBALIZING VILLAGE I: LANDSCAPES OF LABOUR

> Like the rest of the world, the Caviteños live in a rapidly changing society characterized by amazing development and progress. This is the surprising period of technological advancement and miracle of science that empower [sic] all individuals, communities, nation and turn the world into an active global village.
> (Province of Cavite, Promotional Pamphlet, 1997)

The data presented in chapter 3 demonstrated that, among all provinces in the Philippines, Cavite has experienced the most intense processes of urban and industrial development driven by foreign capital investment over the last decade. It is therefore to Cavite that we will turn in this chapter to explore some of the ways in which this experience has been inscribed on the social landscape, and particularly through local processes of labour market formation and regulation. The creation of an industrial labour force goes to the heart of how the people of Cavite have experienced the changes wrought by globalization. The process is experienced both economically, in terms of livelihood and production, and culturally in shifting attitudes to work.

The focus here will not, however, be on the socialization of an agricultural workforce into a factory setting, nor will much attention be paid to the multiplier effects of the industrial sector in the local economy. Both of these themes have been well-covered in literature on the Philippines and elsewhere (Chant and McIlwaine, 1995; Guerrero *et al.*, 1987; Warr, 1984, 1985). Instead, the creation of an industrial workforce will be examined in the local context of a specific cultural ecology, political economy and set of power relations. As a prime agricultural region, the lowland towns of Northern Cavite were, until the 1990s, predominantly rural market centres with rice growing hinterlands. Hence in 1980, 31.2 per cent of the province's workforce was engaged in agriculture, hunting, forestry or fishing. In such a context, a distinct labour process exists embedded in the cultural ecology of rice cultivation. By 1995, the proportion of employment in agriculture had fallen to 8.8 per cent. Meanwhile, the industrial component of the workforce had increased from 11 per cent to 17.3 per cent over the same time period. There has, therefore, been a dramatic industrialization of the province's labour force. This process, along with other labour market changes, including the increasing importance of overseas work and the construction sector, intersects with the farm economy on which most villagers still depend for their livelihood and subsistence. Thus in examining the imprint of globalization on the social landscape, we must consider both the ways in which this industrial workforce has been created and regulated, and how it has integrated with the social and economic organization of agricultural production.

In the first part of this chapter, attention focuses on the provincial scale and the ways in which the local political economy described at the end of chapter 3 constructs a labour market corresponding to the needs, or perceived needs, of foreign capital. The second part of the chapter then explores the relationship between the new industrial labour force and the province's agricultural sector. Here, I present a detailed account of the socioeconomic arrangements behind farming in one village in Cavite, located in the municipality of Tanza, including tenancy, cropping patterns, the labour process in farming, and social divisions of labour within both villages and households. This provides a picture of the social relations of agricultural production which form the context for recent changes to the local labour market. The village in question is called Bunga and while no formal claims of representativeness can be made for the village (its selection was as much serendipity as design, as described in the methodological appendix), I believe the experiences described in Bunga are illustrative of the impacts of labour market changes on rural areas in Northern Cavite as a whole.

While this chapter contains a considerable amount of empirical detail, such detail serves to illustrate several broader points. By looking at the ways in which processes linked to 'globalization' are actually played out through labour processes in a particular setting, three themes emerge. Firstly, globalization as an idea or discourse serves to legitimize and reinforce an existing system of power relations within the localities discussed. Secondly, that system of power relations actually constructs, in very significant ways, what globalization will mean at the scale of human experience. Thirdly, it is through a variety of locally constituted processes, linked to political power, agricultural labour processes, and the cultural dimensions of rural production and reproduction, that processes labelled as globalization are actually played out.

Provincial politics and the social landscape of globalization

In promoting the province of Cavite to foreign investors in the early and mid-1990s, political leaders emphasized the conducive atmosphere for low-cost manufacturing in the province. A promotional brochure, in English, Chinese and Japanese, published in the early 1990s, stressed the attraction of stability and control in the labour process:

> Cavite gives utmost priority to the maintenance of industrial peace and enhancement of productivity in the province's scheme of things. Realizing the importance of these two factors in further speeding up development of their province, peace-loving Caviteños are earnestly helping maintain a climate conducive to business and industries. Thus, Cavite has been tagged as the Zone of Industrial Peace and Productivity, a title coveted by other provinces in the country.

Plate 3 Billboards proclaiming Cavite's transformation

> For the leaders of Cavite, maintaining industrial peace in the province is a 24-hour task . . .
>
> (Province of Cavite, n.d.: 1)

The government's commitment was also exemplified in several campaigns, which found their way onto billboards around the province (see plate 3). One slogan heralded the arrival of 'Cavite's Second Revolution', consisting of 'Industrialization, Urbanization, Tourism Development and Agro-modernization'. Others repeat the declaration of Cavite as an 'Industrial Peace and Productivity Zone', ostensibly by popular acclamation of the 'People of Historic Cavite' in May 1991. While on one level innocuous boosterism, these intrusions in the landscape, directed at both workers and investors, also leave no doubt as to the priorities and determination of the provincial government. Its self-description in terms of 'strong and decisive leadership' and a 'single development oriented political team' further emphasize the coercive force that lies behind such statements (see also, Coronel, 1995; Sidel, 1998)

The application of this power in the process of labour regulation has been extensive. Few aspects of the labour process have escaped governmental regulation, non-regulation, or influence: recruitment, union organization, workplace conditions, pay and benefits, job security and industrial relations. In each case the intention has been to further Cavite's self-proclaimed status as an 'Industrial Peace and

Productivity Zone' and thereby attract industrial investment. Even after his election defeat in 1995, which was widely attributed to the province's 'no-strike' policy, Remulla remained unrepentant in explaining his philosophy:

> Cavite maintains the policy of industrial peace and productivity. In the late '80s and early '90s, when strikes paralyzed industry, Cavite remained peaceful and productive. I, as governor, personally mediated between management and labour to arrive at an amicable resolution of the issues on the basis of what is just, fair and equitable. I practised labor law before I joined politics, and it was as a practising lawyer that I realized how strikes can be disastrous for both management and labor. It was then that I resolved that, given the chance, we should avoid confrontational strikes and resort to amicable settlement. It is not only sound economic policy, it is also socio-culturally appropriate for us non-adversarial Filipinos.
>
> And yes, when militants resorted to violence I was willing to uphold the majesty of law. Those who are not willing to use the potent force of the law when the situation calls for it have no business being called leaders.[1]

At the recruitment stage, there was a widespread belief that preferential treatment was given to those with connections to local political leaders. One such allegation concerned the close links between Remulla and the *Iglesia Ni Cristo*, a Philippine protestant congregation that opposes union membership and is notorious for voting *en masse* in elections for the candidates endorsed by the church's leaders. It has been suggested that members of the church received preferential consideration for jobs in Cavite's industrial estates because of this connection (Coronel, 1995: 17; Sidel, 1995: 382). Municipal mayors and even village captains were also reputed to have influenced the recruitment process through an unofficial system in which an applicant receives the endorsement of a local official. This influence can also be exerted through official channels by granting or denying an applicant the necessary police clearance for employment. Political officials (along with other entrepreneurs) also operate recruitment agencies supplying workers to local factories. This arrangement allows the agency to take a commission from the new employee's wages, and in both official and unofficial placements, to establish a patron–client relationship with the employee for whom employment has been provided – a relationship which, it is assumed, will manifest itself at election time. Such recruiting agencies, some illegal, also operate for the deployment of overseas contract workers.[2]

Much of the influence of local political bosses in the industrial labour process lies in their implicit sanctioning of abuses of labour laws.[3] Regulations concerning health and safety at work are loosely applied, with many reported cases of violations – in one factory in Cavite, for example, it was alleged that at least fifteen workers had lost fingers without proper compensation.[4] Workers also find themselves in positions where they are personally or physically abused by factory managers. The following press report emerged in 1995 about a garment factory in the Cavite Export Processing Zone:

> Workers from a garments firm complained that their Korean manager has been maltreating and overworking them and has even called Filipinos 'easy money'. The Korean manager, a certain M. S. Lee punishes women by ordering them to squat for long periods of time. Male workers are punished by push-ups. The firm is called DAI Young Apparel and is located at the Cavite Export Processing Zone. The workers told Labor Secretary Nieves Confesor over the telephone that toilets are locked during working hours and are only opened during breaktime. Lee and other Korean supervisors in the firm also have a habit of hitting them, they said.[5]

There are also cases of child labour (that is, below the legal working age of sixteen), although parents are usually complicit in this practice since it is often on the basis of falsified records.

Regulations concerning wages and benefits also seem open to abuse. In 1995, the statutory daily minimum wage in the manufacturing sector was 138 pesos (approximately US$5.20 at that time), but firms could employ workers at 98 pesos for a probationary or training period of six months. A widespread practice has been to keep employees at a probationary wage on a permanent basis (in one reported case a firm's employees were still on probationary wages after three years), or, to fire workers and recruit again after six months rather than regularize existing workers and pay them the full minimum wage.[6] The constant supply of fresh recruits needed for this practice is provided by the recruitment agencies often established by local political bosses. There have even been allegations that Governor Remulla was personally being paid a commission from every employee's wage at the Export Processing Zone – with some suggesting amounts between 10 and 20 pesos a day – by companies wishing to ensure the Governor's continued acquiescence to abuses of labour laws. It would seem, however, based on evidence since Remulla was removed from power in 1995, that such accusations were probably apocryphal.[7]

In other cases, wage regulations have been ignored altogether. Reports suggest some companies pay far below even the probationary wage level and some also fail to pay salaries on time and neglect to make the appropriate contributions to the government Social Security System for their employees.[8] In 1995, the Department of Labour and Employment itself estimated that 30 per cent of firms in the country's Export Processing Zones were not complying with minimum wage legislation.[9] It should be added, however, that many firms do adhere to labour regulations and word quickly spreads among workers at the Cavite Export Processing Zone concerning those factories that are the most notorious abusers. Where possible, workers will attempt to switch companies to avoid such practices.

One of the key elements of the province's self-marketing as an 'Industrial Peace and Productivity Zone' has been an unofficial ban on unionization and strikes. Several factors have served to discourage the emergence of an organized labour movement. Firstly, factories have overwhelmingly opted to employ young, single, female workers, who come either from local farming families or as new migrants to the area seeking wages to send home to their families in more impoverished provinces. These young women are generally supervised in the factories by older women and they in turn by male managers – in this way the age and gender-based seniority which exists at home is reproduced in the workplace. Labour unrest is thereby discouraged by the mores of familial discipline. Secondly, the rapid expansion of the industrial sector in the 1990s meant that factory jobs were easily acquired by young women with appropriate qualifications. Thus, as mentioned above, those experiencing dissatisfaction in their working environments are more likely to simply switch companies rather than attempt to change the practices of their current employer through collective action. Thirdly, lessons have been learned from the experience of the earlier export processing zones. In Bataan, for example, the dormitories established in the zone to house workers proved to be fertile ground for union organization. In Cavite, workers have instead been dispersed throughout the neighbouring towns, either in their home villages or in boarding houses. Fourthly, local political leaders exert influence on workers with varying levels of directness, from an unspoken sense that in joining a union a worker would be displaying ingratitude for a recommendation, to the use of direct physical force, which has been alleged in the activities of the province's paramilitary Industrial Security Action Group.[10]

In a variety of ways, then, local political power relations have intersected with the process of industrial development to shape the way in which globalization is experienced. While based on a driving motive of personal enrichment and political entrenchment, this intervention by local political bosses has also sought to supply the workforce needed by foreign investors in the export-oriented manufacturing sector. Foreign capital is courted with assurances of compliant regulatory structures and labour control represented as 'industrial peace and productivity'. Such interventions are, like national development policy, legitimized when the need arises by reference to a globalization discourse in which Cavite is represented as a node in global flows of capital which must be harnessed unless growth is to be sacrificed.

The result has been the creation and regulation of an industrial labour force. Between 1980 and 1995, Cavite's working population (defined as those over fifteen years of age and gainfully employed at any time in the previous year) employed in manufacturing increased from 25,374 to 89,357 (NSO, 1990; NSO, 1997). At the same time, the province is virtually free of union activities; it was not until 1995 that even the mainstream Trade Union Congress of the Philippines established a presence in Cavite with a Labour Education and Counselling Centre near the Cavite Export Processing Zone (Sidel, 1998). To understand the ways in which these transformations in the local labour process have been embedded in social and economic practices, we will now turn to barrio Bunga.

Barrio Bunga and the rural labour process

The village of Bunga lies alongside the Canas River about 5 km from the town proper (*bayan*) of Tanza. Before the war (1941–5 in the Philippines), Bunga still had large areas of forest and uncultivated scrubland.[11] At that time there were just nine houses in the village, but the Japanese occupation of General Trias (commonly known by its old name of Malabon) across the Canas River drove several more families to move to the village to avoid the brutalities of war. One woman in Bunga remembers her mother suffering a miscarriage after a severe beating from Japanese soldiers who suspected her of feeding guerrillas. By the end of the war there were fifteen to sixteen houses in the village and about six extended families. Even now, most of the village is related through birth or marriage to one of these six families.

At that time there was no road to the town, but just an 'alley' (*eskinita*) along which villagers could walk for two hours to reach the *bayan*. It was in 1952 that a dirt road was opened up to link the village with the town. Children would have to walk to school in the town, meaning that many of the older generation did not complete beyond a few years of elementary education. In addition, a small canoe (*banca*) connected the village to General Trias across the Canas River.

In 1975 a horse-drawn carriage service (*kalesa*) started to provide transportation between the village and the town. The major improvement in transport, however, came when the road was asphalted in 1980 and a tricycle service started between Bunga and the *bayan* in 1983. This service is still the principal means of transportation (although by 1995 the road was badly potholed) and allows slightly easier delivery of produce to the market and children to school. With the growing number of young people working at CEPZ in the 1990s, the tricycle service has become more frequent. A further result of improved communications has been that outside buyers and wholesalers now visit Bunga to purchase both rice and vegetables, meaning that transportation to market no longer limits vegetable cultivation. Nevertheless the village is still relatively secluded (*liblib*) from the rest of the municipality.[12]

The village's infrastructure has expanded over the last two decades. Bunga's elementary school was constructed in the 1970s and in the early 1980s electrical power was first supplied to the village. In 1994 a bridge, road and aqueduct were constructed across the Canas River by the National Irrigation Authority to link Bunga with General Trias (see map 4 and plate 4).

Land acquisition and tenancy

For historical and ecological reasons,[13] most farmers in Bunga find themselves farming land that they do not own. For the vast majority their use rights to the land are based on a tenancy relationship.[14] Of the fifty-two farmers that I interviewed in 1995, just five actually owned the land they tilled. At a formal level, tenancy involves paying a percentage of their crop to a landlord as rent (*buwis*). The usual payment would be 25 per cent of their rainy season harvest, remitted as sacks (*cavans*) of unhusked rice (*palay*).[15] More informally, however, the relationship between a farmer and the landlord would usually extend to a 'patron–client' style bonding between two families, wherein, for example, a landlord might act as the godparent to a tenant's child and lend money for medical or educational expenses. This relationship might, furthermore, pass through several generations – tenancies, like ownership, can be inherited by both male and female children. In fact, even during their own working

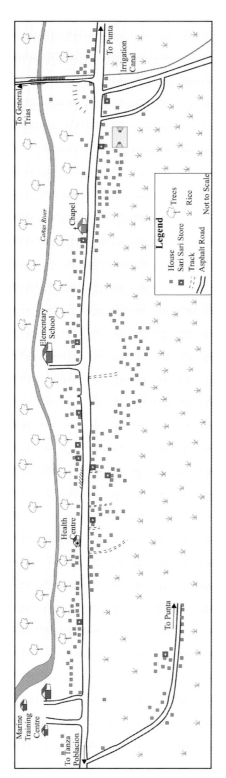

Map 4 Sketch map of barrio Bunga, 1995

Plate 4 The house of an agricultural labourer on the main street through barrio Bunga

life, tenant farmers are essentially free to allow whomever they wish to farm the land, as long as rental payments are made.

Data collected by government agricultural technicians and my own survey data suggest that farmers in Bunga cultivate an average of 2.06 hectares.[16] Although no time-series data are available, information from interviews suggests that there has been substantial subdivision of lands among heirs over the last few decades.[17] This has left many tenant farmers with land holdings insufficient to support a family: thirty-one out of forty-three farmers surveyed in Bunga, or 72 per cent, have two hectares or less.

In just a few cases farmers own the land they cultivate, usually because they have participated in an agrarian reform programme, such as Marcos' Presidential Decree 27 or Aquino's Comprehensive Agrarian Reform Programme starting in 1988.[18] But for the most part, the system of non-cultivating landowners and tenant farmers, with its roots in colonial social structures, still prevails.

There is substantial hearsay evidence that the agrarian reform process has been hindered or corrupted in many instances, thereby preventing tenant farmers from becoming owners of their land. Some farmers believe that landowners have sought to retain ownership by interceding with local government officials and ensuring that Certificates of Land Transfer (CLTs) are stopped.[19] The motive for such an intervention is usually to ensure that landowners will still reap the benefits of conversion at some point in the future. Other farmers, however, prefer to remain as tenants rather than convert to ownership since amortizing their CLT would require onerous cash payments to the Land Bank. Rent, on the other hand can be paid in kind with a share of the harvest.

Without exception, the owners of land in Bunga live in other barrios of Tanza, the town proper of the municipality, or further afield in Cavite and even Manila. Tenants who started to farm in the 1940s or 1950s often did so on the basis of an informal agreement, having either approached the owner, or having themselves been approached with a request to farm the land. Most contemporary farmers, however, have acquired tenancy rights by inheritance (on both parents' sides). The general practice is to pass at least part of a land parcel onto a son or son-in-law on their marriage in order to provide a means of support for the new family. Alternatively, when a farmer gets too old to actively cultivate, unmarried sons may become the *de facto* tenants on the land. In other cases, land will simply be left to some or all children after death. Farm holdings are generally operated by nuclear households. Extended families and children who have left to form their own households do not ordinarily take a share in the harvest. In many cases, however, farming relatives will ensure that their closest relatives do not have to buy rice and will give them what they need for daily consumption, although such generosity is always dependent on their own financial circumstances and their relationship with the particular relative in question.

The farming economy also extends far beyond those who have access to land through a formal tenancy agreement. Many others participate in, and depend upon, the agricultural system. Most are not, however, strictly landless labourers as there exists a variety of levels of access to land. Some will be *de facto* tenants because they are relatives (by birth or marriage) of the official tenant and are waiting to inherit the land. Alternatively, they may be temporarily farming a piece of land through an agreement with the tenant to use it either over the short term, for example to plant one season of a vegetable crop, or over the longer term to plant rice in a system of informal subtenancy. Others are permanent 'helpers' (*katulong*) on a particular piece of land and might work either for wages or a share of the harvest. There are many, however, who simply work wherever and whenever there is farm labour needed.

Just as diverse arrangements might exist between tenants and other agricultural workers, so the relationship between landlord and tenant is also subject to a variety of interpretations. While the statutory rental is 25 per cent of the rice harvest, some landlords will accept much less and only token rentals are usually paid on vegetable harvests. Others, however, will strictly enforce their rights, even to the point of sending observers at harvest time to ensure that an appropriate share is taken and that all debts are repaid. Thus, while the general picture might be simplified as one of (a) absentee landlordism, (b) tenancy and (c) landless labour, there is in fact great variability and flexibility in both access to the land and in the relationship between owner and tenant.

Table 4.1 The cropping cycle in Bunga

Months	Activities
January–February	Dry season (*tagaraw*) rice (*palay*) is planted. Some vegetables planted and harvested
March–April	Dry season rice is harvested
April–May	Vegetables are again cultivated on part of the rice land
May–August	Planting (depending on rainfall) of rainy season (*tagulan*) rice
October–December	Harvesting of rainy season rice

Source: Author's interviews, 1995

Crops and cropping patterns

Traditional farming systems in Cavite allowed for one rice crop per year and some market gardening. Since the mid-1970s, however, new varieties of rice introduced under the 'green revolution' have allowed two harvests per year and higher yields. At the same, the new varieties required increased inputs of fertilizers and synthetic pesticides, which further enhanced yields. The use of machinery, such as handtractors for ploughing, also became more widespread.[20] Farmers in Bunga reported an increase in yields from around 50 *cavans*/ha to around 75 *cavans*/ha. The faster growing new varieties also allow a second harvest during the dry season, although the yield for the second harvest is often considerably reduced (30–40 *cavans*/ha) due to poor water availability even in irrigated areas.

The introduction of new varieties of rice in the mid-1970s was carried out by government agricultural technicians who also trained local farmers in the cultivation of a variety of vegetable and fruit crops.[21] The result was their widespread adoption as farmers noted the substantial profits being made, for example in watermelon cultivation. But the diffusion of such crops was still subject to the marketing limitation imposed by poor transportation in some areas. In Bunga, for example, it was not until the village's road to the town proper was asphalted in the early 1980s and tricycles started to ply the route that vegetables were widely adopted. For contemporary farming in Bunga, table 4.1 shows the usual annual cycle of cropping.

Many farmers will also plant vegetables semi-permanently on small patches of land and, since the area of vegetable crops that can be adequately tended is limited, some farmers may also allow others to plant vegetables on their land between rice crops. This arrangement is variable and could be based on sharecropping, payment of rental, or perhaps simply for free. Some farmers are reluctant, however, to allow their land to be used in this way for fear that the temporary occupant will develop further designs upon the land and attempt to insert themselves as permanent tenants. As a result the usual pattern is for vegetables to be cultivated with family labour alone or with extra labour brought in under a contractual or sharecropping arrangement. Vegetable cultivation can be extremely lucrative if a particular crop has not been widely grown and is consequently selling for high prices in the market. Among farmers rumours abound concerning families in other towns who have made millions of pesos from 'jackpot' harvests of watermelons. But without the benefit of a jackpot, farming families face their leanest months between August and October, when expenses for the rainy season rice crop are mounting, the harvest is not yet in, and rice stored from the dry season crop may already be exhausted.

The agricultural labour process

Despite rumours and aspirations of vegetable 'jackpots', the economic realities of rice farming remain harsh and profit rates are marginal. Based on interviews and survey returns, I will piece together a picture of the typical labour process (and associated balance of accounts) for a notional farm in Tanza. The process starts with land preparation and moves through stages such as planting, fertilizing, harvesting and marketing (all cost estimates are for one hectare and are in 1995 prices).

The first set of expenses are dedicated to land preparation and planting. Water buffalo (*carabao*) are still used as draft animals both in ploughing and in transportation, but ownership of an animal is not universal and during lean times livestock may be sold off to pay for educational, medical or even everyday household expenses (twenty-two of the forty-three farmers surveyed in Bunga owned at least one *carabao*). Ownership of a motorized hand tractor is even rarer but the productivity gains from using one are substantial.[22] One farmer estimated that he could plough in a week an area that would take a month to cover with a *carabao*. As a result, most farmers rent a machine at a rate of 5 *cavans* of rice per day on two occasions during each cropping. In addition, the farmer must purchase fuel for the tractor, costing about 1,000 pesos for the ploughing of one hectare.

Farmers will sometimes keep their own stock of seeds from a previous harvest in order to avoid the expense of purchasing them from a dealer. Frequently, however, such a purchase is necessary because of the need for a new variety of rice. Each harvest with a particular variety of rice diminishes from the one before as plant diseases adapt to the genetically engineered strains. Farmers who are conscious of this process will therefore adopt a new variety every second season. Another reason for purchasing seeds is if the farmer is using a direct seeding technique known as *sabog* where the rice seed is cast directly on the field by hand. The irregular pattern of planting that results makes weeding difficult so farmers try to buy rice varieties which grow with a long stalk that will outgrow the weeds. Until four or five years ago few farmers used *sabog* regularly, but it is now becoming a more widespread practice (the reasons for this trend will be discussed later). If rice seeds are purchased for this or other reasons, it is usually from a dealer and costs the farmer about 400 pesos per *cavan* of seeds (or 8 pesos per kilo). Approximately three *cavans* of seeds are needed on one hectare.

If the usual procedure of transplanting is to be used – as it must be in the more important rainy season rice cropping – a team of planters will be assembled and paid daily (or sometimes hourly) wages by the farmer (see plate 5). Between 1990 and 1995, the standard daily agricultural wage in Bunga increased from 80 to 100 pesos, but some farmers in Bunga reported that labourers coming from other barrios were

Plate 5 Planting rice; but housing developments are encroaching

still asking just 80. In addition, farmers are often expected to provide mid-morning and mid-afternoon snacks for the planters, although demand for such benefits varies. For one hectare of rice land, approximately 15–20 person-days of labour will be needed.

During the growing season, new varieties of rice require inputs of pesticides, herbicides and fertilizers in order to maximize yields. While farmers in dire straits will frequently apply less than the recommended dosage of pesticides and herbicides (known collectively as *gamot* or 'medicine'), estimates of total costs for one cropping range between 700 and 1,200 pesos. Around four sacks of fertilizer, at 350–400 pesos per sack, will also be applied to one hectare.

Farmers are charged for irrigation water by the National Irrigation Authority, although many refuse to pay these dues, complaining that the supply is inadequate and unreliable. The source of the water depends on location. Some parts of Tanza are supplied from the Second Laguna de Bay Irrigation Project, a World Bank funded scheme to bring water westwards from Laguna de Bay into Cavite. Other areas, including Bunga, are irrigated from dams in local streams and rivers. During the rainy season the supply (combined with rain) is more than adequate, but in the dry season farmers frequently face a shortage and in some areas a system known as *manggasahan* is used, whereby a particular area receives water only once during a rotation lasting five years. Consequently, many farmers plant a dry season crop only once every five years and for the remaining time cultivate vegetables on their land. Those farmers who pay irrigation fees reported current costs of 600 pesos in the rainy season and 900 pesos in the dry season for one hectare.

Unlike labour at the planting stage, harvesters are paid a percentage of the rice that they harvest. From the farmer's point of view this encourages a speedy completion of the task, which is vitally important with a window of opportunity between ripening and wastage lasting only a few days. From the harvesters' perspective the share provides food for their families and/or an automatically price-indexed income. The usual share of the yield for harvesters is 10 per cent, but this may increase to 12 per cent if the farmer does not provide snacks for the workers. In cases where the farmer has had to attract labourers from far afield (a trend that will be discussed later), it may also be necessary to provide transportation, food and lodging for the duration of their stay. Approximately ten to fifteen harvesters could complete 1 hectare in a full day's work.

After harvesting, the rice must be threshed to separate the grain from the stalks. This task would once have been done by hand or, according to one farmer, through a traditional technique that involved allowing a *carabao* to trample the crop in a pit lined with its own hardened excrement. The use of a mechanical thresher is now standard, however, and farmers must pay 10 per cent of the harvest to the owner of the machine and the team operating it.

After the deductions for harvesters and threshers are subtracted the farmer must pay rent (*buwis*) to the owner of the land. The traditional *kasama* system of tenancy dictated that the landowner would pay for all of the expenses incurred in farming and then take a 50 per cent share of the harvest. Since the 'green revolution' of the 1970s (and the increased input costs that resulted) the relationship between landlord and tenant has instead become based on a 25 per cent rental payment in the form of *cavans* of rice, with the tenant meeting all operating expenses. In practice, however, this relationship varies and while some landlords will insist on their formal share, others are satisfied with a fixed amount that is sometimes significantly less and does not vary with the yield. One example of the latter type of arrangement in Bunga involved a farmer paying 15 *cavans* per year from 1.5 hectares yielding approximately 180 *cavans* over two harvests.

After threshing and sacking the *palay* in the field, the harvest must be brought to a rice mill. Most barrios with substantial agricultural lands have at least one mill, but farmers must usually pay for transportation from their fields unless they own a jeepney. This expense ranges between three and five pesos for each cavan being transported, depending on the distance involved.

All of these expenses must be met from a harvest that is susceptible to the vagaries of climate, pests and diseases. The official expectation is of 100 *cavans* of *palay* from one hectare of good, irrigated land during the rainy season if it has been appropriately treated with fertilizers and other inputs. In reality, the farmers surveyed in Bunga harvested an average of 56 *cavans* per hectare, largely because cash shortages result in inputs being below the optimum level. One farmer reported using just half of the recommended fertilizer dosage on his land. In the dry season, the total yield is inevitably lower because only land that has reliable water supply is cultivated. But on

Table 4.2 Rainy season cash expense estimates for a 1-hectare rice farm, 1995

Input	Cost (pesos)
Tractor fuel (for two days)	2,000
Seeds (3 sacks @ P400)	1,200
Planters (15 for 1 day)	1,500
Pesticide and Herbicide	1,000
Fertilizer	1,500
Irrigation	600
Transportation	200
Total	8,000

Source: Interviews and survey data, 1995

such land in Bunga the yield was 67 cavans per hectare, reflecting the fact that only the best land is used and that inputs of labour, fertilizer, pesticides etc. can be applied more intensively. If direct seeding, or *sabog*, is used, however, additional weed growth causes the yield to fall to about 35–45 *cavans* per hectare.

The *palay* is sold immediately to a dealer as few farmers have adequate storage facilities to keep either *palay* or milled rice (*bigas*) in commercial quantities. Of the forty-three farmers surveyed in Bunga after the 1994 rainy season harvest, seventeen sold palay to buyers from the neighbouring province of Laguna, ten to buyers from the neighbouring town of General Trias, two to rice millers in other barrios of Tanza, six to local dealers in the village itself, and eight did not sell their surplus crop at all. Farmers will attempt to get the best possible price for their harvest, but the government-regulated prices for rice are frequently ignored by dealers and rice millers. At harvest time, it is inevitably a buyer's market:

> The farmer here has a big problem including those who buy the palay. Even though you know that *bigas* is expensive, *palay* is being bought at a very low price. You as a farmer cannot control the timing of the harvest and you have to sell even at a very low price. You can see the buyer getting rich very quickly. We have a law but if we tell this to the buyer the response is 'Sell it to the newspaper'. So what can we do?
>
> (Farmer in Mulawin, 1995; translated from Tagalog)

The selling price will vary depending on individual negotiations and the quality of the rice, but in 1995, one cavan of *palay* in Bunga sold after the rainy season harvest for an average of 5.5 pesos per kilo or 275 per *cavan*, and after the dry season crop for 7 pesos or 350 per *cavan*. Against this purchase price can be set all of the cash outlays described above for one rainy season harvest. The total costs for one hectare are shown in table 4.2.

This notional figure of 8,000 pesos per hectare compares with an average estimated expenditure per hectare of about 6,318 pesos among surveyed farmers during the

Table 4.3 Rainy season rice account estimates for a 1-hectare farm, 1995 (in *cavans*)

Harvest	70
Minus expenses	
Tractor rental	10
Harvesters (at 10 per cent)	7
Threshers (at 10 per cent)	7
Rent to Landowner	10
Net surplus	36

Source: Various interviews and survey data, 1995

1994 rainy season cropping. The disparity between the two figures probably reflects the fact that not all expenses necessarily grow proportionally as farm size increases above one hectare, and because many farmers are not incurring all of the costs outlined above. Some, for example, were using their own seeds, not paying irrigation fees, not paying for transport and applying less than the recommended dosage of fertilizer and pesticide.

If, however, the inputs in table 4.2 are assumed to have been applied, then a harvest of 70 *cavans* is not unrealistic. Table 4.3 shows the claims that would be made upon that harvest. At 275 pesos per *cavan* in 1995, 36 *cavans* would fetch 9,900 pesos, leaving just 1,900 (US$73 in 1995) after cash expenses have been deducted – the net gain from three months of work.

In fact, most farmers would sooner retain rice for family consumption than sell it all on the open market. In the rainy season of 1994, eight of the forty-three farmers surveyed in Bunga sold none of their harvested rice, and instead retained it all for household consumption. After the dry season crop of 1995, eighteen farmers sold no rice. Of the remaining farmers, most kept a substantial part of their net surplus for domestic consumption. The reason is clear enough: rice is a staple part of every diet and would cost 10–20 pesos per kilo, depending on quality, to purchase in the municipal market.

Evidently, then, rice farming is an economic activity with very slim profit margins and most farmers find themselves short of the capital needed to finance the inputs for a cropping. As a result many take out loans for initial expenses or for other reasons (such as medical or educational expenses or emergency household repairs). Private loans are generally made on the 'five-six' principle, meaning an interest rate of 20 per cent over the period of the loan (usually a few months). In the survey of forty-three farmers in Bunga, over 75 per cent borrowed money to meet their needs for cash during the growing season. The overwhelming majority of these loans, moreover, were made within the village, between private individuals, relatives or the local farming cooperative (see table 4.4).

The result is a situation in which rice farming is only marginally, if at all, profitable. Several farmers described their situation using a metaphor of 'sinking' (*palubog*). Consequently, rice cultivation has, for many farmers, become simply a means of

Table 4.4 Sources of agricultural financing for farmers in Bunga, 1995

Source of Financing	Number of farming households
None	10
Bank	3
Individual outside Bunga	2
Individual in Bunga	13
Relatives	9
Farmers' Cooperative	6

Source: Survey of Farmers in Bunga, 1995

subsistence rather than profit. During interviews I frequently asked farmers why they would not abandon growing rice altogether and focus on more profitable vegetable crops. The response was invariably that rice provides a basic food staple. To many farmers it would be unconscionable to purchase rice in the market and thereby be vulnerable to market forces for the most basic of necessities. But in addition, the social relations embedded in the human ecology of rice (for example, rentals paid with *cavans* of rice) make it more than just another crop to be cultivated. It is a practice of some considerable cultural importance. Thus, most farmers adopt a strategy of growing *palay* for their food needs and relying on vegetable crops to provide for household expenses in the cash economy.

In 1995, thirty-five of forty-three farmers surveyed in Bunga were cultivating vegetables on plots ranging in size from 50 square metres to 3 hectares (in the latter case, the entire farm area was being used for vegetables between rice crops). For most, however, vegetables are cultivated on 300–400 square metres of land on a permanent basis.[23] The costs and benefits involved in vegetable cultivation are far less predictable than rice, but if market prices are high at harvest time, the returns can be considerable. One farmer in Bunga was cultivating string beans (*sitaw*) on a 300 square metre plot and the crop consumed just 400 pesos for fertilizer and 600 for insecticide. All labour inputs were from family members and the yield after harvest was 6,000 pesos. But, as farmers repeatedly told me, vegetable cultivation leaves a great deal to luck or fate (*suwerte*), with respect to climate, water supply, diseases and market conditions.

Another source of occasional income is through livestock rearing. Livestock are owned by thirty-five of forty-three farmers surveyed in Bunga, ranging from a single cow up to a profitable stock of cattle, poultry or swine. For most, however, livestock holdings consist of little more than a few head of cattle and some chickens, which will be consumed or sold in times of need or celebration.

Social divisions of labour

Two divisions of labour run implicitly throughout this account of the economics of rice farming. The first is the social stratification of households within the village according to economic well-being and access to resources. The second is the distribution of responsibilities according to age, gender, marital status, birth order and other culturally constructed categories which determine, in general, how household and productive tasks are assigned to individuals.

Village level stratification is complex and is not easily determined simply from survey information on occupation, which was collected, or income, which was not. Nor can stratification be reduced to the concept of 'class'. While access to land as the means of production is an important factor in determining socioeconomic status, it is far from the only one. With every categorization scheme there inevitably arise overlaps and exceptions. It is, however, important to attempt such a categorization, if only to dispel the myth that the village 'community' in some way represents a group of individuals with a common agenda and identical experiences of change. Based on interviews, surveys and observation, the following categories usefully subdivide the village socioeconomy in Bunga.

Firstly, there are inevitably linkages with those who do not live in the village but exert a strong influence on social relations there. Most importantly, landowners who collect rent on land (and sometimes houselots) from villagers play a role both in this capacity and often as creditors in the agricultural process. In addition, local politicians must be included, given their influence on agrarian reform implementation, infrastructure development, land conversion, and employment.

Within the village, those with cash remittances coming regularly from relatives working abroad are usually among the wealthiest residents, even though their relatives might simply be a maid or construction worker. They will participate as part of a broader class of renters and creditors within the village, owning capital goods such as rice mills, threshers, and hand tractors and lending money to farmers for their operating expenses. In his study of a barrio in Central Luzon, Brian Fegan identifies this class of villagers as the principal beneficiaries of rice cultivation – what he calls 'accumulation on the basis of an unprofitable crop' (1989: 169). Such individuals might also act as small business entrepreneurs or hold village-level government posts. This category will also often include professionals such as engineers and teachers who have no direct relation with the farming economy.

The wealthiest class of tenant farmers might form part of the renter/creditor category, but would also participate directly on their own account in the productive activities of agriculture. Those in a position not to borrow money for their own farming operations and to lend to others are likely to be farming households with a diversified range of activities, including fruit and vegetable crops, flowers, livestock, and perhaps some entrepreneurial sidelines such as a variety (*sari-sari*) store.

The vast majority of farmers, however, are those who need to borrow money for their agricultural activities and yet still manage to make ends meet from a farm of viable size (over about 2 hectares), some livestock, diversified crops and perhaps some money from children working at local factories. In a more tenuous position are those farmers who do not have legal tenancies, but who farm other's land either seasonally or semi-permanently through the variety of subtenancy arrangements described earlier.

Finally, there are households who do not have regular access to land and who earn food and money by working on other people's land on a daily basis, for example at harvest or planting times. These households are by far the most economically marginal and have arrived at their current circumstances for a variety of reasons. Some simply have not inherited any land from their parents who had tenancy rights but too many children among whom to share them. Some are themselves the children of landless parents without the resources to have kept them in school beyond a few elementary grades. Others are migrants to the area and must therefore live with the double jeopardy of being without both land and a kinship network to support them.

These 'classes' are necessarily vague, but provide a preliminary understanding of the economic and social divisions or labour and power within a farming village. Alternative schemes have been proposed based on studies in other parts of the Philippines. Kerkvliet (1990), for example, identifies groups for a Central Luzon village based on the broad categories of 'class' and 'status' distinguished by Max Weber. Four status groups, based on standard of living, emerge in Kerkvliet's study: the very poor, the less poor, the adequate and the rich. Classes meanwhile divide into working class, peasants, petty entrepreneurs, and capitalists, with each category subdividing further. The groups I have identified bear some relation to Kerkvliet's, but combine elements of both class as relation to the means of production and status as standard of living and prestige.

Within the household a division of labour exists that allocates tasks to various members. When asked directly about divisions of labour, many interviewees, male and female, responded that there is no separation of responsibilities and that 'they are just partners' (*parehas lamang sila*) who 'help each other' (*tulungan na lang sila*). Even less direct questions about the generic characteristics of a 'good' wife or a 'good' husband did not elicit clear distinctions in roles – 'you're together, you need to help one another' (*magkasama kayo, dapat magkatutulungan*). Despite these claims, a division of labour clearly does exist. For some activities it is not rigid and allows a variety of household members to participate in an activity while it remains the principal responsibility of one person. In other cases, where divisions are firmer, transgressions occur only in individual cases, and may draw comment and even mockery. For example, interviewees suggested that if a choice has to be made where work needs to be done in the fields but children must be supervised, then it will inevitably be the mother that stays behind.

In many cases, however, it is clear that women carry out numerous agricultural tasks such as watering, weeding, fertilizing, planting, harvesting and marketing produce, as well as acting as financial and commercial managers of the operation. Indeed one woman, whose household engaged in a wide range of farming activities including growing *sampaguita* flowers, keeping ducks for eggs, and growing rice, vegetables and fruits, described (only partly in jest) her husband's role in the following way:

> He's just my sidekick here, I just order him around here, 'water the sampaguita, weed it.' The problem with your weeding is that you will just

Table 4.5 Distribution[a] of household chores in Bunga by gender, 1995

Household chore	Female	Male
Cleaning	64	18
Cooking	54	16
Household repairs	4	40
Childcare	24	0
Marketing	48	11
Laundry	61	19
None	40	59

[a] Number of survey respondents participating in each activity
Source: Author's household survey, 1995

weed the bit at the end, and tomorrow you'll be somewhere else. You'll see him there. It's not connected [i.e. not methodical].
(Farmer in Bunga, 1995; translated from Tagalog)

Most farming activities, such as planting and harvesting, are carried out by both men and women, although sometimes groups will be formed for these purposes that are exclusively male or female. Some interviewees argued that a male-only group reflected the fact that men harvest faster than women and so it would be unfair on other members to admit a slower harvester. Others, however, suggested that women have many other responsibilities, particularly childcare, and so cannot commit to a formal grouping as easily (particularly one that takes life so seriously). A few activities are almost exclusively male preserves, such as threshing and applying chemical pesticides. But as in other activities there are those who transgress identities: 'Here in our place I can see a man's work also being done by a woman' (Farmer in Bunga, 1995; translated from Tagalog).

The precise division of labour between farming couples ultimately depends on the nature of the individual relationship, but there are few activities from which women or men are culturally excluded. In practice, however, women's additional responsibilities in the domestic sphere usually mean that most fieldwork is assigned to men. Recent changes in this division of gender roles will be discussed later in this chapter.

In the domestic sphere most tasks fall primarily to women, although here too men participate to a degree that differs between individual relationships. As table 4.5 shows, women dominate all household activities except household repairs. There are, however, a few cases where husbands stay at home to provide childcare and wash clothes etc. while their wives work at local factories.

Children also engage in farm and household work. Their level of participation depends on age, educational/occupational status and willingness to help. Those below six years old are generally not expected to contribute anything to household or farm work. Older children, however, can contribute to chores outside school time, or on a full time basis if their parents have been unable to keep them in school – a decision that is usually precipitated by financial hardship:

> At present, my third child is in second year high school and is close to a vacation. As a parent it is my dream for her to finish high school. And a parent, even if poor, dreams that she will go to college. But if the parent cannot afford it, then the child has to stop even in high school.
> (Farmer in Bunga, 1995; translated from Tagalog)

The duties that children perform would revolve around minor household chores, looking after livestock and bringing food to workers in the fields. Older children (over ten to twelve years), however, can participate in the full range of farming activities, including harvesting and planting. In both cases younger workers earn the same amount as an adult. For older children who have entered the industrial workforce, however, expectations regarding household and farming chores are usually reduced.

All of this assumes that children are willing to help. While they *can* participate in farming activities or household chores, they are often reluctant to do so. Most parents expressed exasperation at this, but would decline to coerce their children into helping. Much of this attitude towards farming among young people can be attributed to changing youth and work cultures that are themselves closely linked to broader processes of industrialization and urbanization in Cavite. These changes will be examined in more detail later in this chapter.

Reworking the local labour market

It is in this context of agricultural work that globalized development has been occurring in Cavite. The most significant change in the village of Bunga, and many other villages like it, is a transformed local labour market in which farmers must compete with other new economic activities for the services of the workforce (Kelly, 1999c).

For Bunga specifically, the employment opportunities is the Cavite Export Processing Zone (CEPZ) in the neighbouring town of Rosario have had the most far-reaching impacts. The zone first started to attract investment in the late 1980s and by 1995 the CEPZ had a workforce of 38,264, of whom 7,110 lived in Rosario itself and 6,043 in Tanza.[24] In addition, approximately 800 more residents of Tanza worked in the three factories established in the town itself. Most of these workers are between the ages of sixteen and thirty. Meanwhile, over the last few decades, many others have taken jobs overseas, usually as domestic helpers in East Asia, Europe or North America, construction workers in the Middle East, or merchant seamen. The result has been significant changes in the local labour market as a predominantly agricultural town in Manila's rural hinterland has become swept up in changes resulting from the rapid industrialization of the 1990s described in chapter 3.

A large literature exists on the subject of non-farm employment in rural areas of Asia. Much of this work assumes, however, that non-farm work is still centred on the farming family, for example in household craft production, rather than full-scale factory-based industrialization (Islam, 1987). A literature has also emerged that examines the socialization of rural workforce into a 'modern' factory regime (Ong, 1987; Pinches and Lakha, 1992; Wolf, 1992). Little, however, has been written of the impact of factory work on agricultural activities as mediated through the local labour market. Nevertheless, the development plans envisaging globalized development for Cavite assume that farm and factory production will coexist harmoniously. The key word throughout is 'balanced' development. A major objective of the Calabarzon Masterplan, for example, is to 'sustain high levels of growth on *the balance between agriculture and industry* by promoting their complementary linkages, improving the industrial structure and inducing related service activities' (JICA, 1991: 4). The Cavite Provincial Development Plan strikes a similar tone and provides a land use plan based on 'the assumed development thrusts of the province, namely, *industrialization, agricultural modernization, tourism development, and rapid urbanization . . . integrating urban functions to that of agricultural development*' (Province of Cavite, 1990: 50; emphasis added). Even the Tanza municipal development plan proposes that 'the municipality should promote and establish a self-sustaining economic structure within the context of *a balanced agro-industrial type of development*' (Municipality of Tanza, 1995: 109: emphasis added).

What none of these documents addresses is the way in which industrial development actually intersects with the experiences of farming households. But the export-oriented industries that have developed in Cavite draw, to a large extent, on the same local labour force as existing agricultural activities. In the rest of this chapter, I will examine the ways in which globalized development has intersected with farming activities through the transformation of the local labour market. Bunga will provide the case study for this examination.

The sectoral structure of the village labour market

According to the 1990 census of population, 1,103 people were living in Bunga in 205 households. My own survey in 1995, covering 230 of the 260 houses in the village, recorded 1,154 people, which by extrapolation would suggest a total population of approximately 1,300. This represents an annual increase of around three per cent, which roughly corresponds to the national rate of population growth over the same period. Map 4 shows the contemporary settlement geography of this population. The age and sex distribution of the population is given in figure 4.1. There is little that is unexpected in the village's population profile, but figure 4.1 will serve as a useful basis for comparison with a second village case study in the next chapter.

The occupational structure of the village is provided in table 4.6. While this table provides primary and secondary household incomes, it was evident during qualitative interviews that sources of livelihood often extend to a diverse range of activities. The concept of 'a job' or 'a career', which is easily assumed in designing survey instruments from a Western perspective, clearly did not apply. While only two occupations have been included here, the fact that the total number of secondary occupations is almost

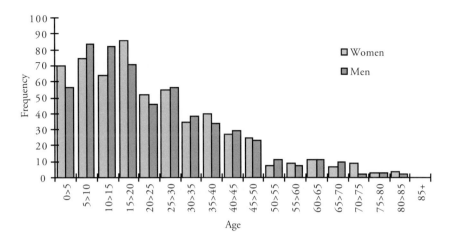

Figure 4.1 Age and Sex structure of Bunga, 1995
Source: Household surveys, Bunga, 1995

as high as primary occupations strongly implies the prevalence of livelihood diversity. To give an indication of the overall economic structure of the village's economy, these figures are aggregated in table 4.7 and figure 4.2. The data demonstrate the essentially agrarian nature of the village economy, and the importance of factory employment as a secondary income source.

Overseas work

As noted in chapter 2, the Philippines has been a major exporter of labour since the early 1970s. In Cavite there were 32,623 registered overseas workers in 1995, of whom 85 per cent held high school graduation diplomas or higher qualifications (NSO, 1997). The differentiation by gender contradicts the popular image of a predominantly feminized process; 71 per cent of these workers were men.

Like most other Filipino villagers, people in Bunga aspire to the foreign currency earnings of working abroad and the financial security that it can provide in a local context. Relatively few households have had such an opportunity, but the benefits to those that have are immediately apparent. Solidly built houses, electrical appliances and comfortable furnishings all indicate that someone in the household has been, or is currently, working overseas. In Bunga, 25 out of 231 surveyed households included someone who had worked abroad in the past. A further 16 households currently had contributing members working abroad (2 households had 2 people, giving a total of 18 individuals currently abroad). Tables 4.8 and 4.9 provide a breakdown of this total by destination, gender and occupation.

Although the numbers involved in overseas work remain relatively small – currently removing only eighteen individuals from the labour market – its effect is felt

Table 4.6 Primary and secondary household occupation by gender, Bunga, 1995

Sector/occupation	Primary		Secondary	
	Female	Male	Female	Male
Agricultural/resource based				
Tenant farmer or farmer with land	2	60	1	0
Farmer/helper on a relative's land	0	15	1	6
Farmer attached to other's land	0	5	0	0
Farm labourer	10	26	14	16
Vegetable planter on other's land	2	16	0	4
Thresher operator	0	2	0	1
Palay dealer	0	2	2	0
Cultivator and seller of sampaguita flowers	1	0	10	2
Catching fish	0	1	0	0
Rice miller	0	0	0	1
Caring for other's livestock	0	0	0	2
Keeping livestock	0	0	3	6
Native cheese vendor	0	0	1	0
Vegetable vendor in market	0	0	1	0
Factory work				
Factory worker at CEPZ	6	2	51	4
Factory worker elsewhere	0	2	5	1
Government work				
Philippine National Police officer	0	1	0	0
Garbage collector	0	0	0	1
Government employee	0	0	2	0
Elected government official	0	0	0	2
Armed forces: military/civilian	0	4	0	0
Teacher	1	0	2	0
Construction and related sectors				
Construction/contract worker overseas	0	1	1	0
Carpenter	0	5	0	0
Construction worker	0	12	0	8
Electrician/mechanic/welder	0	2	0	2
Engineer	0	1	1	0
Mason	0	1	0	0
Painter	0	1	0	0
Maintenance worker	0	2	0	1
Real estate agent	1	1	0	0
Retail/distribution				
Delivery boy	0	2	0	0
Fish vendor	1	0	4	1
Salesperson	0	1	0	0
Sari-sari storekeeper	3	1	12	0
Restaurant waiter/dishwasher	0	1	0	1
Snack vendor	0	0	2	0
Department store assistant	0	0	1	0
Junk dealer	0	4	0	2
Transportation and services				
Domestic helper	2	0	9	0
Dressmaker/home sewer	1	0	2	0

Table 4.6 (cont.)

Sector/occupation	Primary		Secondary	
	Female	Male	Female	Male
Driver	0	5	0	0
Gambling coordintor	0	0	1	0
Labandera (laundry)	0	0	6	0
Tricycle driver	0	10	0	7
Secretary	0	0	1	0
Seaman	0	2	0	1
Security guard	0	2	0	0
Midwife	0	0	2	0
Nanny/babysitter	0	0	2	0
Medical/dental assistant	1	0	0	0
Barber	0	1	0	0
Computer Programmer	0	1	0	0
Total	31	193	137	69

Source: Household surveys, Bunga, 1995

Table 4.7 Total occupational structure of Bunga, by sector, 1995

Sector	Female	Male	Total	% of working population
Working population				
Agricultural/resources	48	165	213	49.8
Manufacturing	62	9	71	16.6
Government	5	8	13	3.0
Construction & related sectors	3	36	39	9.1
Retail/distribution	23	13	36	8.4
Transportation/services	27	29	56	13.1
Non-working population				
Dependency	3	0	3	n.a.
None/Student	411	308	719	n.a.

Source: Household surveys, Bunga, 1995

beyond this number alone. For every person that is abroad, or who has previously been abroad, there is one person, or perhaps more, who need not engage in the gruelling work involved in harvesting or planting rice because of the 'dollar' income that is coming in. In addition, the possibility of working abroad has an effect on the aspirations and work attitudes of the younger generation (Kelly, 1999c). Thus, in one case, a farmer's son could refuse to work for his father while he did nothing but wait for his papers to come through to become a seaman.

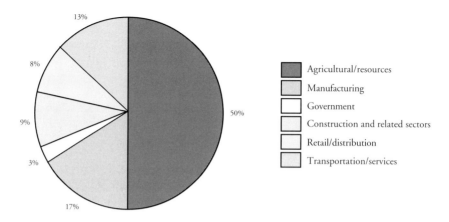

Figure 4.2 Occupational structure of working population in Bunga, 1995
Source: Household surveys, Bunga, 1995

Table 4.8 Currently deployed overseas workers from Bunga, by destination, 1995

Location of work	Male	Female	Total
Australia	1	2	3
Europe	0	2	2
North America	1	1	2
Middle East	6	1	7
Japan	2	1	3
Taiwan	0	1	1
Total	10	8	18

Source: Household surveys, Bunga, 1995

Table 4.9 Currently deployed overseas workers from Bunga, by occupation, 1995

Occupational Type	Male	Female	Total
Construction	1	0	1
Computer Programmer	1	0	1
Domestic Helper	0	7	7
Driver	1	0	1
Electrician	1	0	1
Factory Worker	2	1	3
Seaman	3	0	3
Unknown	1	0	1
Total	10	8	18

Source: Household surveys, Bunga, 1995

Plate 6 The CEPZ: a banner entices factory workers to the even richer rewards of the global economy by working overseas

Factory work

For many, overseas work is unavailable or undesirable, and so for the majority of people in Bunga the main point of contact, in economic terms, with a globalized economy is through employment at a factory in the CEPZ (see plate 6). Finding work there is known to be relatively easy. Job vacancies are usually filled through word of mouth as workers tell their friends and relatives of openings who then approach company managers in person. Although the broader literature on the labour process in Cavite, particularly in the publications of progressive groups, suggest that political connections are the key to employment, only a few workers living in Bunga secured their jobs with this kind of help. More important was to have a 'backer' inside the factory itself who could vouch for the applicant and act as '*padrino*' (literally godfather, but in this case a 'sponsor' or 'go-between').

Individual firms vary in their recruitment practices, with garment factories known to have less stringent requirements than electronics companies. A minimal process would involve the completion of an application form and an interview either with supervisors (usually Filipino) or factory managers (usually foreign nationals). Some firms may also administer an exam and require a letter of recommendation from a *barangay* captain, councillor or the municipal mayor. The documentation required includes a birth certificate, police clearance, *barangay* clearance, residency certificate, Social Security System papers and educational transcripts. The clearances are intended to assure the employer that the worker has no criminal record and has a fixed address in a local village.

While these are the formal mechanisms for recruitment, there also appear to be other selection criteria at work in filling vacancies at the CEPZ. According to the law, a person must be sixteen years of age to hold full time employment. Many, however, falsify their age and are taken much younger, including thirteen year olds. Villagers believed that factories turned a blind eye to this practice because they prefer to take younger workers who will be more malleable and socialized into a parental mode of discipline that can be transferred to the factory floor. Single and childless individuals are also seen to be preferred, leading some women to conceal their marriage and motherhood from factory recruiters. The unstated ceiling for employment appears to be around 40 years.

Educational status also appears to be a sensitive issue for recruiters. While garments factories will accept workers with only elementary school graduation, many in the electronics sector require high school completion, and, for supervisory positions, college graduation. At the same time, interviewees in Bunga felt that the desire for well-qualified workers was limited:

> There are those who were able to work at EPZA and yet didn't finish schooling. They just borrow a diploma. They don't like bright people at EPZA. There was someone from Santol [a neigbouring village] who was very intelligent and yet wasn't accepted. She [or he] was studying law. Why? Intelligent people know about unions. They don't like unions there at EPZA. Those people know the ins and outs of [a union].
> (Farmer in Bunga, 1995; translated from Tagalog)

The most significant filter on the employment process, however, appears to be based on gender. As a whole, 77 per cent of employees at CEPZ are women (see table 3.4 in chapter 3), but in Bunga the proportion is 90 per cent. It seems likely that selection on the basis of gender aims to draw upon the same culturally-rooted power relations as the age-bias. Women are socially constructed as being under the protection of men, with role-models for girls providing images of nurturing wives and mothers. Such constructions, fostered at home and school, are employed on the factory floor where supervisors are generally older women and managers are men.

One young man expressed his perception of the situation in the following terms:

> It's really easy for women to be accepted. It seems they trust women, but men they don't seem to trust. Because I think men must be very bad [said sarcastically].... Maybe it's on trust. Companies trust women much more than they trust men. Because men really fight back, women do not, they are soft-hearted [or weak].
> (Young man, Bunga, 1995; translated from Tagalog)

The typical profile of a line worker at CEPZ then, is a young single woman with at least some high school education. Middle aged male farmers and farm labourers do not fit the profile for employment at the CEPZ. 'I don't fit there' (*Hindi na ako maaari doon*), one commented, reflecting on his age and lack of formal education.

Many factories run shifts around the clock and so employees find themselves working for eight hours in time slots that can vary from month to month. Thus while many women working at CEPZ have ambitions to continue studying, it is difficult to plan any sort of further education around the variable working hours at the factory. Most workers get to CEPZ by catching a tricycle from Bunga to the main highway in Tanza and from there a jeepney or bus to the gates of the zone.

Workers are paid every two weeks either through direct deposit to a bank account or in cash, with deductions for Social Security contributions. In 1995 the legal minimum daily wage in the manufacturing sector was 138 pesos after deductions, but companies can legally pay 95–120 pesos during a probationary period lasting six months. These regulations are adhered to by some companies, but others breach them flagrantly. In one case, a probationary worker was started at 25 pesos a day, and wages of 80 pesos were not uncommon. In some cases, companies also failed to pay wages on time or neglected to make appropriate Social Security contributions.

Working conditions in factories varied. Garment factories are generally acknowledged as being harder work than electronics firms, as the work often involves standing for hours at a time. Some workers also complained about the coldness of the air-conditioned factory environment. The working week is six or seven days, although some factories will lay off workers temporarily if there is a period of low orders. Shifts are eight hours long with thirty minutes for lunch and two fifteen minute breaks.

The compensation for this work, if the minimum wage is paid, amounts to about 2,000 pesos every two weeks (equivalent to approximately US$77 in 1995). From this must be subtracted the costs of transportation and food:

> EPZA helps a little. The children that have work can help. . . . But they can't save, because, firstly, the income is small, it's not large there. The expenses with that work, for transportation a large amount is deducted, and the food – they're eating in the canteen [at the factory]. Almost nothing is left to them.
>
> (Parent in Bunga, 1995; translated from Tagalog)

Many in Bunga observed that the salary is barely sufficient for one person, and workers at the CEPZ are often in a position of having to borrow money from their parents for transportation. Yet many workers are also able to pass on a part of their salary to their parents.

The sharing of income from CEPZ with parents is usually voluntary, but undoubtedly expected. The amount contributed ranges from a few hundred pesos to half or even the whole salary, in which case parents will provide them with an allowance. In farming families, the additional cash income can preclude the need for borrowing to cover farming expenses. Two parents reflected on the benefits of employment at CEPZ:

> Wife: It's a big benefit because they're working and they're giving money.
> Husband: EPZA is also a big help for people from the barrio. Young people who don't have jobs just go and work there at EPZA, helping the parents.
> (Couple in Bunga, 1995; translated from Tagalog)

Some interviewees compared the relative contributions of sons and daughters.

> Sometimes, like my daughter who works at a factory, this last season, she's the one who helped me with the planting. My daughters help. That's why I can say that daughters are really better [than sons]. She gave me money for expenses. I'm not really asking her. For example, [I will say] 'Neneng, we'll be planting soon, and I have no money, I'll have to borrow it'. She gives it to me, voluntarily.
> (Farmer in Bunga; translated from Tagalog)

In another case, a farmer compares his son and his daughter. For his son, aged twenty-two, who lives at home and works as an agricultural labourer: 'if his parents don't ask he will not give' (*kung hindi hihingi ang magulang, hindi magbibigay*). But for his daughter, aged nineteen, who works at a CEPZ garments factory, 'she's giving, just like that, naturally as she's living in my place' (*nagbibigay, katulad nga niyan, siyempre dito siya nakatira sa akin*). In addition to contributing to the family expenses, working children also cover their own costs of living:

> That's a big help for us parents, because whenever she lacks money, if we spend for her it's just a little. It's not like when you're with your parents and you don't have a job. Your every move, every time you eat, and your clothes – all will be shouldered by your parents.
> (Farmer in Bunga, 1995; translated from Tagalog)

One could be left with the impression that these young women are in a situation of double exploitation – working for low wages in a factory and then persuaded by gentle moral coercion to turn over their earnings to parents at home. But daughters too expressed their willingness to contribute and actually spoke of their contribution as enhancing their status in the household, 'because, actually, if you are earning a salary, you can help a little' (*kasi, actually, makakasweldo ka na iyon parang nakakatulong-tulong ka na*) (Female factory worker in Bunga, 1995). The share of the income kept for themselves can also be spent on consuming items that would otherwise be unaffordable – for example, trips to go 'malling' in Manila and buying clothes and fast food.

The changing culture of work

As urbanization and industrialization proceed, cultural changes have occurred that create a younger generation with very different value orientations, career horizons and ambitions from their parents. One area of particular significance to this discussion is the reformulation of meanings imbued in work. Several related dimensions of work culture are pertinent here: the changing attitudes of farmers towards their profession, from self-respect to self-denigration; the changing attitudes of young people towards agricultural work, which directly affects the pool of labour available at harvesting and planting times; and, the changing aspirations of young people more generally, away from agrarian occupations and towards the opportunities presented by factory work, city work and overseas employment (see Kelly, 1999c).

As other opportunities have presented themselves, young people have shifted their aspirations and expectations away from a rural life and towards other forms of work. As farming has become less appealing, its place is being taken by the greater economic and cultural capital to be accumulated from factory work, and even more so from office/retail jobs in Manila or overseas work. A youth leader in Bunga commented on how few of his generation have any interest in developing and applying the farming skills of their parents. The same youth leader also noted that even where young adults don't have the sort of work to which they aspire – for example if they are waiting for a posting as a seaman – they will often still prefer not to help in agricultural tasks. The labour shortages experienced by farmers are not, therefore, purely the result of potential workers being employed elsewhere, but also reflect a change in 'attitude' (*ugali*) towards the type of work that is desirable.

Needless to say, the older generation take a rather dim view of such changes. A farmer in Bunga had the following to say about the changing cultural proclivities of the younger generation:

> There is a great change among the youth of today compared with before. It's because now the young people are too social [*sosyal*]. Now the young people are becoming more 'elegant' [*magagara*] than before. Because in those days [when I was growing up] you won't believe it, I never experienced wearing long pants.
> (Farmer in Bunga, 1995; translated from Tagalog)

Two key words in this quote warrant some elaboration. Literally, *sosyal* translates as 'social', but in this context it is being used in a more pejorative manner. The implied meaning is behaviour in an elite or aloof manner, or with an attitude ('to take on airs' might be an appropriate English equivalent). The implication is that younger people have become socialized into a higher way of life and consider themselves too 'precious' to be involved in farm work. A similar implication is inferred by the verb *magagara*, whose root, *gara*, literally translates as elegant. A more accurate rendition might, however, be pompous, extravagant, fancy or dressy.[25]

A commonly cited indicator of unwelcome cultural change is the growing affiliation of adolescents to *barkadas*. In general usage, the word *barkada* refers to a peer group of friends and can relate to men or women of any age. In one of the few explicit considerations of the phenomenon, Jean-Paul Dumont defines the male *barkadas* he encountered in the following way:

males, first as boys, then as men and well into their old age, belong to and participate actively in such informal, but class-bound and long-lasting gendered groups of coevals. . . . In the process, not only were long-lasting bonds created between these males but male stereotypes were thoroughly reinforced.

(1993: 402–3)

The literal translation of '*barkada*', as a group of passengers together on a boat, implies something of its figurative meaning – a group of peers, usually from adolescence, embarking on parallel journeys through life. But, as Dumont points out, the adoption of the word in Tagalog only emerged in the 1950s and came with a connotation of rebellion or mischief.[26] Now, the concept retains that anti-establishment sense, as an egalitarian grouping of men, women, or classmates of both sexes who share a bond of common experiences and among whom social etiquette and inhibitions might be dispensed with or at least lowered. The *barkada* therefore represents a social context in which behaviour can stretch conventional norms – a crucible for redefining the aspirations and identities of youth, and a controlled rebellion against the overbearing institutions of family, lawfulness and hard work. The element of control is found in the conservative aspects of the *barkada*. The group can, Dumont argues, apply peer pressure to subtly foster social and moral conformity among its members and discourage originality and initiative. Nevertheless, for many young people, the *barkada* serves as

> an egalitarian refuge where they could feel less restricted than usual in their actions, less guarded in expressing their feelings and altogether freer than anywhere else. In a way, this was conducive to some exploratory behaviour that co-members of their *barkada*, ever solicitous with each other, both tolerated and chided on an acceptably reciprocal basis.
>
> (Dumont, 1993: 429)

While *barkadas* exist for men and, to a lesser extent women, of all ages, it is among adolescents and young adults that they are most formative and influential. It is in the context of the *barkada* that youth subcultures develop and new attitudes and identities are forged. The youthful *barkada* is widely seen as an unproductive influence on young people by their parents. One farmer in Bunga described how his son would help him with farm work until he was fifteen years old. From then onwards his attention was always elsewhere: 'it was different when he formed his *barkada*, now [he does] nothing, he's becoming lazy' (1995; translated from Tagalog). Another farmer in Bunga spoke of 'attitude' among young people more generally, and the 'vices' they were developing:

> It's difficult for me to [understand] . . . you know people tend to get hooked on vices, and there are vices coming to affect young people, many vices are emerging now. If you're not aware, there are those who are discreetly

grouping themselves, like a fraternity. It's very different now, nothing good will come of it. Sometimes they're smoking up, smoking weed. That's what they usually do. That's why they establish such a grouping. There are many lazy people here, really many. There are really very many lazy young people in our place, and sometimes even if their house is falling apart, they won't help their parents.

(1995; translated from Tagalog)

While this perception of *barkadas* as foci of vice and degeneration is probably overstated, it is undoubtedly true that changes in the culture of work that have taken place in Bunga further exacerbate the problems of labour supply in agriculture. Thus the local impact of urban and industrial development is not limited to the economic sphere. Significant sociocultural changes also occur, particularly amongst the younger generation. Furthermore these changes exacerbate the already tight agricultural labour market by reformulating the cultural capital ascribed to non-agricultural activities (this connection is elaborated in Kelly, 1999c).

Responses to labour market changes

The emergence of overseas and factory work has resulted in substantial changes in the local labour market in Bunga. The pool of agricultural labour that was once available to be tapped at harvest and planting time is now absorbed by other activities:

From when factories first started to multiply in Cavite, many people didn't work in farming, like harvesting, many don't like to harvest, like planters they have disappeared. That's why it's difficult for farmers. That's a big loss.

(Farmer in Bunga, 1995; translated from Tagalog)

A variety of responses to these circumstances has emerged. Many farmers have adapted production techniques to use less labour power, migrant agricultural labour is being widely used, harvesting teams have developed in the village, farming households have increased the use of family labour in a few instances, and cultivation of labour-intensive vegetable crops have been limited. These responses are described in the following sections.

Adaptive production techniques

Labour market changes in recent years have precipitated adaptation in the organization of agricultural work. The planting and harvesting stages of the production process, in particular, have seen important changes.

A common reaction to labour shortage has been the increased use of a direct seeding technique, or *sabog*. This hand-casting of seeds onto the field precludes the need for labour in transplanting seedlings from nursery fields. The random dispersal of seeds, however, does mean that fields are more difficult to weed than a crop planted

Table 4.10 The emergence of *sabog* utilization in Bunga, 1975–95

Year when sabog was first used	Number of farmers
1975–9	1
1980–4	4
1985–9	1
1990	3
1991	2
1992	6
1993	3
1994	6
1995	4
Subtotal who have used *sabog*	30
Not using *sabog*	13
Total	43

Source: Farming survey, Bunga, 1995

in uniform rows, and so yields are typically less. One farmer estimated a decline in yield from around 60 *cavans* per hectare with manual planters to only 35–40 *cavans* using *sabog*. Other farmers broadly confirmed this magnitude of yield reduction. Another disadvantage encountered in using *sabog* is that seeds are highly sensitive to water conditions and are prone to rot if water-logged. In general, therefore, *sabog* can only be used in the *tagaraw* (dry season) cropping, when water from irrigation canals can be closely controlled.

Despite these drawbacks, the use of *sabog* has become increasingly common in the face of labour market changes: 'if we don't have planters, we use *sabog* instead. That's our way, we are then free from expenses. There are many who are doing it' (Farmer in Bunga, 1995; translated from Tagalog). Since 1990, in particular, when employment at local factories started to proliferate, the use of *sabog* has become widespread, as the data in table 4.10 indicate. These figures do not imply, however, that thirty farmers are now always using *sabog* during the dry season cropping. Farmers will decide according to the apparent availability of labour and their own financial circumstances whether the use of direct seeding is economically rational. In the 1995 dry season cropping, for example, fifteen of the forty-three farmers surveyed used *sabog*. What is clear, however, is that its use has increased sharply during the 1990s as a response to changing labour market conditions, with reduced crop yields as a direct consequence.

While the use of direct seeding provides a response to labour shortage at the planting stage, such short cuts are difficult at harvest time. The shortage of agricultural labour has made the harvest a tense time for farmers. A window of only about five days exists between the rice fully ripening and the crop being ruined. It is essential, therefore, to have access to the services of harvesters at the appropriate time. Before recent changes in the local labour market, a sufficient pool of labour existed within each village to meet the needs of its farmers. Now, in conditions of labour shortage, those harvesters that do still work locally have begun to develop teams in order to carry out the work more efficiently. Rather than try to scrape together sufficient labour, a farmer can simply approach a team – either local or migrant – and secure their services for a specific period: 'We formed the group for a reason. Because it's now the trend here. If you're not with a group then you cannot harvest. Farmers like to harvest the palay all at once, that's why we formed the group' (Agricultural worker, Bunga, 1995; translated from Tagalog). The formation of harvesting groups is beneficial for both farmer and harvester. The farmer is provided with an efficient labour supply when it is needed, and the harvesters can group together to work faster and harvest a greater area. Membership of a group is not, however, open to all. Some groups are all male, and even where they are mixed, women may be unable to join because domestic responsibilities mean that they cannot provide a full-time commitment to the group.

Migrant agricultural labour

A second consequence of the changing local labour market has been the migration of agricultural workers into Tanza. In some cases, migration is temporary, as workers from other provinces, particularly Batangas, are transported to Cavite and stay only for the duration of the harvesting period. The arrangements for this circular migration are informal but well established. A representative of a team of harvesters will visit Tanza and assess the likely peak harvesting time for crops already in the ground. He (such groups are, to my knowledge, exclusively male) will then make arrangements with several local farmers to return at the appropriate time with a team of harvesters. The farmer will pay for their passage from Batangas and will also provide accommodation (usually a thatched *nipa* hut near the fields) and meals for the duration of their stay. Alternatively, a farmer might be forced to seek out harvesters either from among teams brought by other farmers or by actually travelling to towns in Cavite or Batangas to find them. In either case, the farmer incurs additional expenses in the form of transportation and board and lodging for the harvesting team.

Migrants from further afield, particularly the Visayas, have opted to stay semipermanently in Tanza, living either on the farmland of their employer if they work for an individual farmer, or in small 'ghettoes' that have emerged in several of Tanza's rural villages. While the reason for selecting a particular village to settle in may be a family connection, however tenuous, the broader structural reason for migrating to the area is the availability of agricultural and other work opportunities. Tables 4.11 and 4.12 show the place of birth, year of arrival and occupational characteristics of migrants to Bunga. The data confirm several qualitative impressions about the nature of migration to the area. Firstly, migration into Bunga has accelerated over the last three decades and figures for the five years between 1990–5 suggest continued growth in the immigrant population. Secondly, the most important source of migrants into the area is the Visayan region of the Philippines – more important even

Table 4.11 Birthplace and arrival date of migrants in Bunga, 1995

| Year of arrival | Region of origin | | | | |
	Southern Tagalog & Central Luzon	Metro Manila	Northern Luzon	Visayas	Total
1950–9	0	0	1	0	1
1960–9	1	0	0	1	2
1970–9	4	3	0	8	15
1980–9	9	7	1	20	37
1990–5	6	11	0	15	32
Unknown	9	4	0	1	14
Total	29	25	2	45	101

Source: Household survey, Bunga, 1995

Table 4.12 Birthplace and occupation of migrants in Bunga, 1995

| Occupational sector | Region of origin | | | | |
	Southern Tagalog & Central Luzon	Metro Manila	Northern Luzon	Visayas	Total
Agricultural/ resource based	10	4	0	10	24
Factory work	1	0	0	1	2
Government sctor	0	1	0	1	2
Construction and related sectors	1	3	0	9	13
Retail and distribution	3	2	0	2	7
Transport and Services	3	0	0	2	5
None	11	15	2	20	48
Total	29	25	2	45	101

Source: Household survey, Bunga, 1995

that Manila and the adjacent regions of Southern Tagalog and Central Luzon. Finally, for those immigrants in the workforce, the agricultural and construction sectors are clearly the most important.

The data in tables 4.11 and 4.12 suggest that just under 10 per cent of Bunga's population was born outside of Cavite. In relative terms this is a small proportion (as will be apparent in the next chapter which describes recent changes in the nearby barrio of Mulawin). Nevertheless, people in Bunga still speak of 'the Visayans' in a way that suggested their presence, even on account of marriage or other relationships, does not go unnoticed and that they are identified as different. While many expressed tolerance towards their new barrio-mates, and several migrants said that they had experienced no problems integrating with the local community, social tensions do occasionally arise. One Visayan woman had the following to say about the reception she had received in Bunga:

They belittle [or look down upon] us Visayans. Because they don't like . . . what I mean is, they really did it to me. When I was very new here. They really belittle us Visayans, here in Bunga. But it's only here in Bunga that I've experienced it. In other places like Manila, Visayans are respected, in other places too such as Olongapo. It's only here in Bunga that they belittle us. They look at us [as if we're] so small. They will say, oh, they're just Visayans.

(1995; translated from Tagalog)

Changing social divisions of labour

A further consequence of labour shortage is an increased dependence, in some cases, on household labour:

> EPZA has a big impact on us. We don't have harvesters now. Even planters, none also. When it's rainy season, the palay often rots because there are no harvesters . . .
> Q: What do people in Bunga do about that?
> They themselves are the ones harvesting, the owner. The family harvests. The whole family. We're the ones harvesting.
>
> (Farmer in Bunga, 1995; translated from Tagalog)

This farmer's statement did not, however, hold true for many other households. Both survey returns and qualitative interviews suggested that household divisions of labour remain intact, not least because it would require an exceptionally large and energetic family to replace the labour input of hired workers. In addition, the changing attitudes towards agricultural work among the younger generation, described earlier, leave farmers exasperated at their children's unwillingness to help in farming tasks.

A further consequence of industrial development is rooted in the predominance of young women in the industrial workforce. Women are favoured in the CEPZ factories precisely because the dominant construction of gender roles places women's formal work at the margins of the economy. Female wage labour is represented as a secondary form of livelihood, even where it may yield a higher income for the household. This denigration of female labour provides the commonly noted paradox found in Filipino gender relations: while women have access to nearly all spheres of economic and political life, their role in these spheres never completely removes the expectation that they conform to models of the caring mother, dedicated wife and obedient daughter. Participation in the workforce is usually a supplement to these duties rather than a substitute. According to the dominant construction of femininity, women are not expected to be the primary bread-winners for the household. Instead, their role is within the household. Nevertheless, the employment of young women in the industrial sector has forced some changes upon existing divisions of labour based on familial hierarchies and gender and broadened the roles which women in marriages or daughters living with parents can adopt for themselves.

There are signs, then, that in Bunga the increased participation of women in the waged workforce is bringing changes in the social construction of femininity and the expectations attached to it. Even while traditional roles retain their currency, and economic relationships might be exploitative, women in the workforce nevertheless benefit from enhanced status. Where daughters are paying for their siblings' education, saving for their own further education, contributing to the household budget or assisting in covering the capital outlay involved in farming, they accrue both respect and a debt of gratitude (*utang na loob*) from their families. This is the reason that many young women commented that they were pleased to be providing financial assistance.

The intangible benefits of added familial respect and status for young factory employees also translates into additional influence over decision-making within the household and often exemption from household chores. Furthermore, waged employment improves the quality of young women's leisure time, allowing for greater indulgence in activities such as 'malling' in Manila, purchasing clothes, and going to movies. In these ways, then, waged employment can be seen as an enhancement of the social position of women by expanding the cultural space in which their self-identity and familial relationships are constructed.

While this enhancement of women's positions in social power relations is not necessarily a 'zero sum' situation, there is clearly some readjustment necessary on the part of men. A young man in Bunga commented, somewhat wistfully, that 'it is now usual, among young married people, for the man to stay at home, and for the woman to work' (1995; translated from Tagalog). In fact, this statement is far from accurate, and most women from Bunga who are working at CEPZ are unmarried and still living with their parents, but it does belie a sense that gender roles are changing. To some extent the same is true in agricultural work, and some male farmers noted the increased participation of female labour in farming: 'here in our place I can see a man's work also being done by a women' (Farmer in Bunga, 1995; translated from Tagalog).

The participation of women in harvesting, for example, is common in Bunga. Yet harvesting teams coming from the town of Lemery in the neighbouring province of Batangas to work in Tanza commented on their surprise at finding women working in the fields. In their own villages, harvesting was an exclusively male preserve. While this evidence is anecdotal, it points to the manner in which gender roles can vary geographically depending on labour market conditions, and suggests that shifts over time are a response to changing circumstances. Such evidence points, therefore, to a shifting of gendered and familial divisions of labour that cuts across both the industrial and agricultural sectors.

Cropping decisions

The shortage of agricultural workers also places limitations on the extent to which farmers can take advantage of the profitability of vegetable cultivation. Even if help could be found, hiring waged labour at current rates would substantially offset the profits of cultivation. Some farmers also expressed reluctance to hire workers for vegetable plots because the constant and careful attention that they require leave them vulnerable to workers who are either less than conscientious or decide to pursue other opportunities in the middle of a cropping. The result is that most farmers limit their vegetable cultivation to an area that they can maintain with their own labour and that of family members. This appears to be at least part of the answer to a question I asked farmers repeatedly during interviews. Why would they not cultivate a greater area of vegetables when such crops were clearly more profitable than rice? The rest of the answer would seem to lie in a cultural attachment to rice cultivation, and the embeddedness of certain social relationships – most notably between landlords and tenants – in the process of rice farming.

At a broader level, these limitations suggest that the supposed intensification of agriculture with urbanization that orthodox theory and regional planners would predict – the agro-modernization heralded on provincial government billboards – is not happening in Bunga, precisely because of the labour market changes associated with urbanization and industrialization. While farmers are growing more vegetables than in the past, due to increased accessibility and technical assistance, expansion of such activities is limited by both socio-cultural factors and by the shortage of wage labour in agriculture.

Conclusions

This chapter has suggested that, at two levels, processes of globalization are embedded in local political, economic and ecological relations. In both cases it is through the labour process that globalized development is mediated and experienced in the social landscape. At the *provincial* scale, the labour process is closely regulated by political power brokers in order to ensure a workforce is supplied that satisfies the perceived needs of foreign capital. At the *village* level, a complex labour process exists and includes social and gendered divisions embedded in the human ecology of rice farming. This context fundamentally affects the ways in which globalization is activated in the lives of men and women in a village like Bunga.

The cultivation of rice remains at the core of the economy for a barrio such as Bunga. Even though a detailed accounting of the household economics of rice farming indicates its marginality, it persists because it provides a key dietary staple and forms a central axis for the social and cultural relationships that structure rural life. The nature of Bunga's economy means that access to land is a fundamental axis of social differentiation. For historical reasons the predominant model of access is through a tenancy with an absent landowner. Rental payments are remitted to the owner after each rainy season cropping in the form of sacks of rice. After this and other expenses have been met, the profitability of rice farming is very marginal. But farmers persist in cultivating rice because it is the basis of their access to the land, which can also be used for more profitable crops such as vegetables. Rice cultivation also continues because it provides the dietary staple of farmers and their families. For many farming households rice is now as much a subsistence activity as it a commercial operation. These are the economic rationalities for rice farming, but the cultural significance of the practice should also be noted. It defines the rhythm of rural life and the social structure of the village.

The experience of globalization in Bunga takes two principal forms: the expansion of overseas work, and, more significantly, the growth in local factory employment at the nearby Cavite Export Processing Zone in the first half of the 1990s. But these opportunities are only selectively available. Just over 16 per cent of Bunga's working population are employed in local factories and such opportunities are certainly not on offer to middle-aged farmers with little formal education. Instead, factory work is predominantly directed towards young single women. This reflects the techniques of labour regulation imposed in many export factories, which are predicated upon local constructions of gender. But in agricultural activities the gendered division of work is more equally shared, with women active in many farming tasks. This is particularly true at labour-intensive times such as harvesting and planting. Similarly, while young men might not be official tenants, and only some can expect to inherit the tenancy rights to a piece of land, they have in the past been counted on to provide farm labour at planting and harvest times. Their withdrawal from the agricultural workforce to take jobs in construction, local factories or overseas, further exacerbates labour scarcity.

The consequence is that farmers are left scrambling to find the labour that they need for cultivation, and paying a higher price for it. Labour intensive rice cultivation persists for the reasons mentioned earlier but adaptations have to be made. Increasingly, this means using alternative farming techniques such as direct seeding, or *sabog*, and a few depend more heavily on family labour. Harvesting teams have also formed to increase the efficiency of the process, and migrant agricultural workers from other provinces – notably Batangas – have also been drawn upon to meet labour needs. Some will concentrate a little more on profitable vegetable crops, but for most this is limited to an area that they themselves can manage. Some local practices have been adjusted, for example through the formation of harvesting teams, but the fact remains that the labour-intensive cultivation of rice is a central part of rural social relations and so labour shortages are experienced.

The incomes earned in factory employment, construction or overseas work can serve as a subsidy to farming operations, for example in eliminating the need to borrow. But while supporting cultivation in this way, the shortage of agricultural labour is at the same time cutting away at the economic and social sustainability of agriculture. For example, where the *sabog* technique is used in rice planting, it translates directly into reduced yields.

Two conclusions, then, can be drawn concerning the consequences of globalized development in Bunga. Firstly, the socioeconomic effects of globalization, most notably the changes in local labour markets, cannot be understood except through the complex social relations that exist in a province such as Cavite or in a rice-growing village such as Bunga. Access to land, social and gendered divisions of labour, and the labour demand cycles of rice cultivation are all key relations, constituted locally, that shape experiences of globalization. Secondly, the consequences of globalization, far from fostering a 'balanced' form of development in which agriculture and industry co-exist harmoniously, instead appear to be undermining agriculture through the changes caused in the local labour market. Thus, even in a predominantly agricultural village such as Bunga, globalized development must be viewed as a political choice to prioritize industrial growth *over* support for improvements in agrarian productivity, rather than a Panglossian move towards balanced agricultural and industrial development.

5

THE GLOBALIZING VILLAGE II: THE PHYSICAL LANDSCAPE

We have examined the ways in which the local political economy of power relations has shaped the creation of an industrial labour force and how the resulting transformation of the local labour market has intersected with an existing labour system in Bunga's agricultural sector. In Bunga, however, the physical landscape of a rural village remains essentially intact. The village has not (yet) seen the widespread land conversion from rice fields to urban or industrial uses that has occurred elsewhere in Cavite. Five kilometres of unpaved road appears to have deterred property speculators and developers thus far. A few kilometres away, however, across a creek and some gently undulating rice land, the village of Mulawin lies on the National Road between Tanza and the provincial capital of Trece Martires City. There, large swathes of land have been taken out of agricultural production. Little farmland in the village has escaped the attention of speculators and developers, and by 1995 only twelve farmers were still cultivating crops (compared with twenty-five in 1989).[1]

Such land has either been converted to residential estates or left idle as tenant farmers are removed and owners speculate on rising land prices. In Mulawin, then, significant changes have been occurring in both the social fabric and physical landscape of the village. The driving force behind these changes has been the globalized development most clearly manifested in the nearby Cavite Export Processing Zone.

The purpose of this chapter is to explore the ways in which these changes occur through local social and political structures and a specific environmental context. In particular, the relationship between landowners and tenant farmers creates a set of social circumstances in which land conversion is judged to be economically rational and socially feasible. The local structures of political power create an environment in which bureaucratic regulations relating to land conversion can be manipulated or circumscribed. This provides a key mechanism through which local political leaders can both foster rapid development through ensuring land availability and, in many instances, can also profit personally from interceding in such transactions. As the process has continued, the ecology of rice cultivation is found to conflict with the environmental changes brought about by urbanization, thereby creating an impetus for all remaining land to be converted. This chapter, therefore, provides some evidence of the locally embedded nature of globalization as a material process inscribed upon the landscape of Cavite.

The first section of the chapter examines the broader process of land conversion in Cavite and suggests that it is significantly more widespread that official national figures suggest. Through various means, the conversion of agricultural land can be facilitated even where it contravenes the spirit or the letter of the law. The following section introduces the historical development and socioeconomic characteristics of Mulawin, highlighting its very different social and economic composition from the village of Bunga. We then examine the nature of property development in the village. Who is building residential subdivisions? What type of environment do they create? Who lives there? In the fourth section, I will describe the context of land conversion in terms of local social and political relations between landowners, developers, tenant farmers, and politicians. The fifth section identifies some of the areas of environmental incompatibility between urban and agricultural land uses and the consequent problems facing farmers – problems that add to the already immense pressures to sell their tenancy rights. The chapter concludes with an assessment of the limitations placed upon agriculture in a context where crops can be grown profitably, and agricultural land is legally protected, but more powerful forces dictate that the land be used for other purposes.

Land conversion in Cavite

The conversion of farmland has been a point of considerable debate within the Philippines, particularly among progressive groups concerned with the eviction of tenant farmers and issues of food security.[2] The process in Cavite has its roots in the suburban expansion of the southern boundary of Manila and in the relocation of squatter settlements from the urban core of the National Capital Region to new towns in the East of the province (see McAndrew, 1994). In the 1990s, however, the process has been driven by the development of large industrial estates in a zone from East to West across the centre of the province (see plate 7). Housing estates have also begun to sprout up to accommodate factory workers and commuters working in Manila (see plate 8).

This real estate boom in Cavite is evident in the population classified as urban for census purposes. The level of urbanization increased from one half to three-quarters between 1970 and 1990 (NSO, 1993). The process is also reflected in dramatic increases in the cost of land in Cavite. In the province as a whole, the cost of home lots increased by an average of 20.4 per cent every year between 1990 and 1993 (this compares with a 5.5 per cent increase in Metro Manila).[3] The conversion of farmland into industrial estates or residential subdivisions has been widespread, but in many instances land has also been taken out of production and tenant-farmers removed as a speculative tactic pending future sale of the land (see plate 9).

Tables 5.1 and 5.2 indicate the land use pattern in 1988, and the subsequent conversions that have been registered with the provincial government. Clearly, residential and industrial developments have been the major causes of land conversion, but there are reasons to suspect that these figures from the Department of Agrarian Reform (DAR) represent only a fraction of the land actually taken out of agricultural

THE GLOBALIZING VILLAGE II

Plate 7 A new factory development in the midst of prime agricultural land

Plate 8 A residential subdivision in barrio Mulawin: most plots remain undeveloped as supply outstrips demand

THE PHYSICAL LANDSCAPE

Plate 9 Rice land now lies idle, awaiting development

production. The owners of land that is lying idle, and therefore not technically converted, have usually paid off tenant farmers in order to avoid agrarian reform redistribution and are waiting for a more profitable moment at which to convert the land to other uses.[4] But many lands have simply been converted without the knowledge of the DAR, usually because grounds have been established for exempting the conversion application from DAR jurisdiction (such as proving a non-agricultural land use or by invoking the municipal land use plan). Still more land has simply been converted illegally without the necessary clearances.[5]

Data from the municipality of Tanza indicates that while 367.3 hectares of agricultural land was approved, exempted or being processed for conversion between 1989 and 1994, a further 221.8 hectares had been converted without permission – in both cases the vast majority being for residential subdivisions.[6] Data on rice lands in the municipality, meanwhile, gathered by local Department of Agriculture extension workers, suggests that over 400 hectares of land conversion was on irrigated rice land.[7] If these figures can be extrapolated – and field observations suggest that Tanza is not atypical of lowland towns in the province – the implication is that official conversion figures must be scaled up by at least 50 per cent to account for illegal conversions, and that around two-thirds of conversion is on irrigated rice lands. This underestimation by official figures would run still higher if unreported or unnoticed illegal conversions were also factored in.

Numerous pieces of legislation exist to protect agricultural land from conversion. The land to be converted must be officially zoned as non-agricultural and it must be

Table 5.1 Land use in Cavite, 1988

Land use category	Area (hectares)	% of total area
Agricultural	106,080.12	74.33
Non-agricultural	36,625.88	25.42
Built-up	21,999.73	15.42
Woodland	13,101.70	9.18
Wetland	1,542.45	1.07
Total	142,706.00	100.00

Source: Province of Cavite, 1990

Table 5.2 Land conversions in Cavite approved or being processed by DAR, 1988–95

	Residential	Industrial	Institutional	Commercial	Farmlot	Tourism	Unknown	Total
1988	7.0				21.0			28.0
1989	17.4	25.0						42.4
1990	159.8	160.8	14.8	0.4				335.8
1991	73.0	35.0	148.7					256.7
1992	209.5	266.6	12.4		67.8		53.8	610.1
1993	99.9	24.8	10.0		12.0			146.7
1994	125.3	7.9						133.2
1995[a]	286.6							286.6
Being processed	98.4					66.0	103.2	267.6
Total	1076.9	520.1	185.9	0.4	100.8	66.0	157.0	2,107.1

[a] Until 30 June
Source: Unpublished Data, DAR, 1995

non-irrigated.[8] It must also not be eligible for redistribution from owner to tenant under agrarian reform legislation. But each of these conditions can frequently be sidestepped. The Local Government Code of 1991 (a piece of administrative decentralization seen as a key component of economic liberalization) enables local officials to reclassify up to 15 per cent of land use from agricultural to other uses on the basis of vague conditions relating to its viability for agriculture. Many towns, however, do not even have precisely defined or publicly available zoning maps, leaving decisions over reclassification to local officials, particularly mayors. Such rezoning decisions often involve bribery and kickbacks.[9] Certification that land is unirrigated is equally open to bureaucratic corruption and other means of circumvention. Areas designated for Regional Agri-Industrial Centres, Tourism Development Areas or socialized housing can be prioritized for land conversion with presidential authority.[10] Alternatively, landowners may deliberately vandalize or neglect irrigation canals and dikes in order to make a case that the land is unirrigated and unproductive.

Clearances from the Department of Agrarian Reform that the land has 'ceased to be economically feasible and sound for agricultural purposes'[11] and that farmers on the land have been properly compensated are also open to abuse both by local officials and landowners. Numerous documented examples exist of pressure being brought to bear on farmers in Cavite who have resisted the decision to convert farmland.[12] Farmers' rights as agrarian reform beneficiaries have also been compromised where redistribution has been prevented or withdrawn, often with pressure being applied to local agrarian reform officials. It should also be said, however, that many farmers have been only too happy to sell their tenancy rights given the marginal profitability of rice cultivation and the often generous compensation packages which are negotiated.

As a result of these various legal loopholes and extra-legal activities, municipal politicians have been able to exert considerable control over the land conversion process (Kelly, 1998). The provincial authorities technically have little power to influence land conversion decisions, but in Cavite there have been widely circulated accusations of complicity on the part of Remulla's government (Sidel, 1995). The principal ways in which this influence has been exercised are through the persuasion or coercion of uncooperative municipal officials and tenant farmers, supplying money to add a further incentive for compliance, and the provision of manpower for forced evictions:[13]

> In numerous documented cases, he [Remulla] has dispatched armed goons, ordered the bulldozing of homes, and engineered the destruction of irrigation canals, so as to expedite the departure of 'squatters' and tenant farmers demanding compensation for their removal from lands designated for sale to Manila-based or foreign companies for 'development' into industrial estates. Though Remulla typically tempers such hardball tactics with offers for a 'settlement, the 'carrot' is never as impressive as the 'stick'.
> (Sidel, 1995: 381)

In the case of the Cavite Export Processing Zone in Rosario, the national government was also involved. When President Marcos declared 275 hectares in two municipalities to be the site of the CEPZ in 1980 (under Presidential Decrees 1980 and 2017) it was then prime irrigated rice land under cultivation. Contractors employed by the Export Processing Zones Authority and the provincial government started to bulldoze the site in March 1981 but faced opposition farmers who organized themselves into the *Samahang Magsasaka at Mamumuwisan ng Cavite* (Association of Farmers and Leaseholders of Cavite) (McAndrew, 1994: 125). A Supreme Court ruling in August 1981 halted construction on the site, but was overturned when the government produced an unpublished Presidential Decree dated seven months earlier (PD 1818) that inhibited the courts from interfering with government development projects. As a result, construction continued and most farmers succumbed to pressure and accepted a compensation package.

The exercise of social power relations also occurs at a personal level. As noted in the previous chapter, the relationship between a tenant and a landowner goes beyond a formal economic contract and encompasses sociocultural hierarchies of class, status

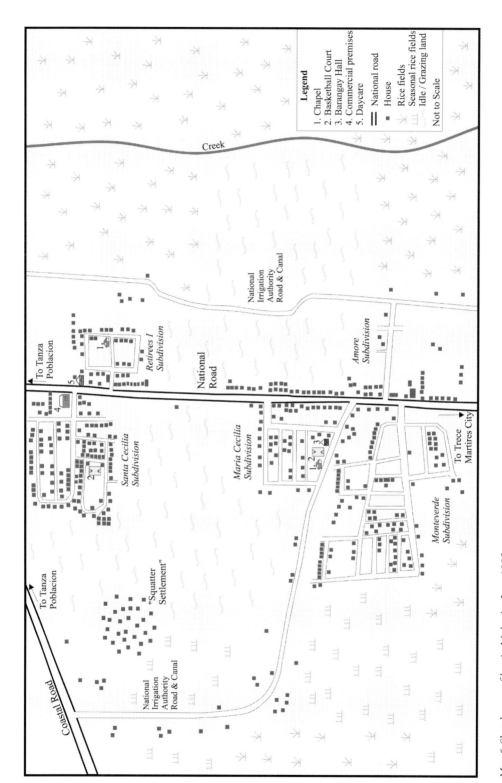

Map 5 Sketch map of barrio Mulawin, June 1995

and clientalism. As a result, tenants are often reluctant to enforce their legal rights or to bargain aggressively for adequate compensation. Thus, the personal context of land conversion cannot be ignored. This will be examined in more detail later in this chapter.

There is therefore a complex process of political struggle occurring over the release of agricultural land for urban and industrial development. This struggle occurs at three levels. At the national scale legislation has created numerous loopholes and opportunities for evasion without adequate deterrents. At the same time, the government's national development strategy has aggressively promoted industrial development while the agricultural sector has been implicitly neglected. At the local level, power over land use decisions has been decentralized into contexts where boundaries between public sector regulation and private economic interests are blurred. This has, in turn, created a political framework conducive to conversion. Finally, at the level of personal relationships between landlords and tenants, the 'everyday politics' of conversion is played out in a cultural context of patron–client ties that preclude farmers from asserting their legal rights.

Introducing Mulawin

Mulawin lies along the cemented national road linking Tanza with the provincial capital in Trece Martires City (see map 5). Its relative accessibility, compared for example to a village such as Bunga, has resulted in an experience of globalized development that is striking in both the social and physical landscape of the barrio. Indeed, the village owes its very existence largely to the recent boom in industrial employment and property development.

Until 1986, Mulawin was a *sitio*, or neighbourhood, of the adjacent barrio of Sanja Mayor, rather than a *barangay* in its own right. In the 1940s, just a handful of extended families occupied the area, living in ten to fifteen houses alongside the unsurfaced road that led towards the central and southern uplands of the Cavite. The area that would later become Trece Martires City was then the remote barrio of Quintana (also in Tanza) and a stronghold for leftist Huk rebels and the New People's Army (NPA). In the 1950s, however, Delfin Montano, a native of Tanza and son of a prominent senator, became provincial governor and moved his administration from Cavite City to the newly created Trece Martires City. The road was asphalted at that time but *sitio* Mulawin was still populated by just a few families.

It was in the late 1960s that the first substantial growth occurred in the barrio, with the construction of the Maria Cecilia residential subdivision on rice land alongside the National Road, followed by the Santa Cecilia subdivision completed in 1972, and Retirees I in 1980 (see map 5). Electricity was first supplied to the area in the late 1970s. Other major housing projects have been developed over recent years, including Amore (1990) and Monteverde (1992), spurred on by the cementing of the national road in 1989. Six residential subdivisions now occupy much of Mulawin's land area, along with vacant land being held as speculative investments by owners or developers. Farmland only remains in areas set well back from the road.

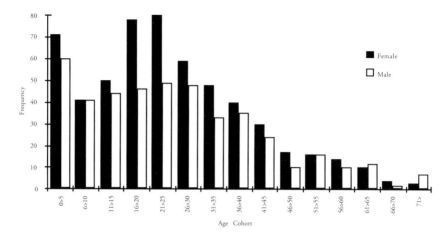

Figure 5.1 Age and sex structure of Mulawin, 1995
Source: Household surveys, Mulawin 1995

By 1986, the population had grown sufficiently to justify designating Mulawin as a *barangay*, and the 1990 census recorded 1,830 people in 341 households. My own survey covered 1,047 people in 208 households – approximately one-third of the barrio's total number of households in 1995. On the basis of this survey I estimate the barrio's total population at approximately 2,750 in 550 households by 1995 – an increase of 50 per cent over five years.[14] This trend, driven by new migration to the village and facilitated by recent housing developments, seems likely to continue as the empty lots in existing subdivisions are filled and new developments are constructed.

Figure 5.1 shows the age and sex structure of the barrio's surveyed population in 1995, and provides a contrast to the relatively balanced population structure of Bunga. Three prominent features of the village's socioeconomy are evident. Firstly, the large numbers of women in the 16–25 age groups emphasizes the importance of employment at CEPZ for local residents, many of whom are migrants to Mulawin. Secondly, the number of both men and women in their late twenties, coupled with the very large number of infants aged 0–5 years highlights the number of young families who have moved into the area to take advantage of employment opportunities and new housing. Thirdly, the discrepancy between male and female numbers in almost all adult age categories highlights the fact that employment opportunities for men often involve leaving the barrio, and in many cases finding work overseas. These features are corroborated by data on the employment structure of the village.

Table 5.3 provides a detailed listing of the occupational structure of Mulawin's household economy. Before commenting on the overall nature of the village's economy, it is necessary to add that unlike Bunga, Mulawin's population does not consist solely of nuclear family/household units, but also includes a number of residents who live in shared rental accommodation, as independent tenants rather

Table 5.3 Primary and secondary household occupation by gender, Mulawin, 1995

Sector/occupation	Primary		Secondary	
	Female	Male	Female	Male
Agricultural/resource based				
Tenant farmer or farmer with land	1	9	0	1
Former tenant farmer	0	7	0	2
Farmer/helper on a relative's land	0	1	0	0
Vegetable planter on others' land	0	0	1	0
Farmer attached to other's land	0	1	0	0
Farm labourer	0	1	0	0
Vegetable seller	1	0	2	0
Pig rearing	0	2	1	0
Palay dealer	1	0	2	1
Catching fish	0	1	0	1
Factory work				
Factory worker at CEPZ	18	15	45	18
Factory worker elsewhere	2	7	5	2
Government work				
Government employee	4	2	2	2
Armed forces or police: military/civilian	1	2	1	1
Elected government official	0	0	0	1
Teacher	1	1	8	0
Construction and related sectors				
Carpenter	0	16	0	5
Construction worker	0	13	0	7
Draftsman	0	1	0	0
Electrician/mechanic/welder	0	14	0	0
Engineer	0	2	0	2
Mason	0	3	0	0
Painter	0	1	0	0
Maintenance worker	0	2	0	3
Retail/distribution				
Salesperson	4	1	4	2
Sari-sari storekeeper	4	0	18	1
Import-Export business	0	0	1	0
Snack vendor	1	0	4	0
Transportation and services				
Domestic helper	1	0	3	0
Dressmaker/home sewer	1	0	0	1
Driver (car/jeep/taxi)	0	15	0	1
Laundry/labandera	1	0	4	0
Beautician	0	0	1	0
Pastor	0	1	0	0
Tricycle driver	0	8	0	7
Television producer	0	0	1	1
Real estate agent	3	1	0	0
Medical/dental assistant	0	0	4	0
Rental income	0	1	1	0
Seaman	0	6	0	1
Restaurant waiter/dishwasher	0	0	0	1
Flight attendant	0	0	1	0
Secretary/office employee	4	1	3	1
Entertainer	0	0	1	0
Shoe repair	0	1	0	0
Computer programmer	0	1	0	0
Total	48	137	110	62

Source: Household surveys, Mulawin, 1995

Table 5.4 Independent residents' occupations, Mulawin, 1995

Occupation	Female	Male
Factory worker at CEPZ	48	3
Live-in domestic helper	12	0
Construction worker	0	6
Carpenter	0	1
Secretarial/office employee	2	0
Tricycle driver	0	1
Total	62	11

Source: Household surveys, Mulawin, 1995

Table 5.5 Total occupational structure of Mulawin, by sector, 1995

Sector	Female	Male	Total	% of working population
Working population				
Agricultural/resources	9	27	36	8.3
Manufacturing	115	45	160	37.2
Government	17	9	26	6.0
Construction and related sectors	0	76	76	17.7
Retail/distribution	36	4	40	9.3
Transportation/services	31	49	80	18.6
Live-in domestic helper	12	0	12	2.7
Non-working population				
Dependency	4	3	7	
Student	111	111	222	
None	241	145	386	
Total	576	469	1,045	

Source: Household surveys, Mulawin, 1995

than households. Known locally as 'bedspacers', and providing a source of rental income for some local households, these individuals are overwhelmingly employed at the export processing zone, as shown in table 5.4. If these independent residents are added to the primary and secondary livelihoods of other households, then an overall picture of the village's occupational structure can be drawn, as provided in table 5.5 and figure 5.2. Two caveats should be noted with regard to these employment data. Firstly, they are constrained by the nature of the survey instrument used, which allowed for only one occupation to be listed for each household member.[15] In reality, this misrepresents the diversity of livelihood activities, or 'sidelines' as they are known locally, that many individuals undertake. One important omission, for example, is the widespread practice of renting land from tenant farmers in order to plant vegetable crops after the rainy season rice harvest (in the period between October and March).

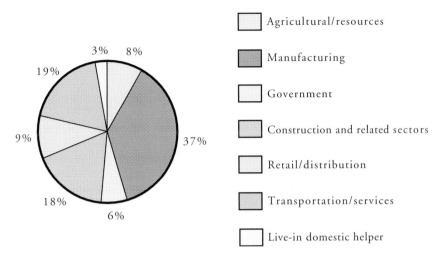

Figure 5.2 Occupational structure of working population in Mulawin, 1995
Source: Household surveys, Mulawin, 1995

In-depth interviews suggested that approximately fifteen to twenty individuals are engaged in agricultural activities in this way, many of them farmers who have sold their tenancy rights but continue to apply their skill on land rented or borrowed for individual croppings.

A second caveat concerns systematic exclusions from the survey's coverage. Logistical reasons limited the survey to 208 households in the core of the village around where the original settlement was located. The survey thus covered the old *sitio* of Mulawin, an established subdivision (Maria Cecilia), a new subdivision (Monteverde) and several areas of fields where some farming households live. Two established subdivisions (Santa Cecilia and Retirees I) were excluded together with a concentration of houses squatting on land to the West of the village proper (see map 5). The latter area contains a large number of migrant workers who are employed in a variety of agricultural activities such as farm helpers and rice mill attendants. Thus it is very likely that the tables above under-represent those engaged in farming on the margins of the agricultural economy.

What does emerge clearly in table 5.5, however, is the predominance of manufacturing employment in the village's economy. More than one third of the workforce is employed in factories, and all but a handful of those are located in the Cavite Export Processing Zone in Rosario. It has been such employment opportunities that have driven the process of land conversion in Mulawin and other barrios in Tanza and that have attracted migrants to the area from other parts of the country. A second major source of local livelihoods, for young men in particular, is the local construction industry, including ancillary sectors such as carpentry, painting, and masonry.

Table 5.6 Birthplace and arrival date of non-Cavitenos in Mulawin, 1995

Arrival year	Place of birth						Total
	Southern Tagalog or Central Luzon	Metro Manila	Northern Luzon	Visayas	Mindanao	Europe	
1950–9	0	0	0	0	0	0	0
1960–9	1	1	0	1	0	0	3
1970–9	7	14	0	7	0	0	28
1980–9	10	29	1	6	0	0	46
1990–5	81	59	19	54	22	2	237
Unknown	2	7	0	3	1	0	13
Total	101	110	20	71	23	2	327

Source: Household survey, Mulawin, 1995

Table 5.7 Birthplace and occupation of non-Cavitenos in Mulawin, 1995

Occupational sector	Region of Origin						Total
	Southern Tagalog or Central Luzon	Metro Manila	Northern Luzon	Visayas	Mindanao	Europe	
Agricultural/ resource based	4	2	0	1	0	0	7
Factory work	35	11	3	16	8	1	74
Government sector	1	3	0	4	2	0	10
Construction and related sectors	8	11	0	10	1	0	30
Retail and distribution	3	3	1	3	1	0	11
Transport and services	13	12	2	8	0	0	35
None	38	68	14	28	10	1	159
Total	102	110	20	70	22	2	326

Source: Household survey, Mulawin, 1995

Mulawin and migratory flows

The employment opportunities described above have attracted migrants to Mulawin, and as table 5.6 indicates, the years 1990–5 in particular have seen a major influx of new arrivals. Data from my 1995 survey shows that 31 per cent of the population at that time were born outside Cavite and approximately 23 per cent of the population had arrived within the last five years.[16] The main sources of migrants are other provinces in the Southern Tagalog and Central Luzon regions (i.e. the national core region to the north and south of Manila), and out-migrants from Manila itself. As in Bunga, Visayans are also strongly represented. Table 5.7 confirms that a major source of employment for such migrants has been in local factories.[17]

Table 5.8 Overseas workers, by occupation and gender, Mulawin, 1995

Occupation	Female	Male	Total
Construction and related	0	6	6
Entertainer	1	0	1
Domestic helper	3	0	3
Driver	0	4	4
Electrician/engineer	0	5	5
Factory worker	2	0	2
Flight attendant	1	0	1
Medical assistant	1	0	1
Office/secretarial	0	1	1
Restaurant worker	0	1	1
Salesperson	1	2	3
Seaman	0	7	7
Unknown	2	2	4
Total	11	28	39

Source: Household survey, Mulawin, 1995

Table 5.9 Overseas workers, by workplace and gender, Mulawin, 1995

	Female	Male	Total
Europe	0	1	1
North America	1	1	2
Middle East	1	15	16
Guam	1	3	4
Hong Kong	2	0	2
Japan	2	3	5
South Korea	1	0	1
Malaysia	0	1	1
Taiwan	1	0	1
Unknown	2	4	6
Total	11	28	39

Source: Household survey, Mulawin, 1995

Another important component of the village economy is temporary migration for overseas employment (see tables 5.8 and 5.9). Of the surveyed working age population (15 years and above) of 740, 39 people were working abroad and a further 44 had done so in the past. These individuals thus represent 72 households from a survey of 208 (34.6 per cent), who have at some time benefited from an income remitted from abroad.

One feature of the data in tables 5.8 and 5.9 that is worth noting – and this was also true in Bunga – is the predominance of men among overseas workers. Perhaps because of the visibility of Filipina domestic helpers and nannies in the West and across Asia, and some recent political controversies over their treatment, there is a popular perception that Filipino overseas migration is a largely female phenomenon. In Mulawin, however, it is men who are most likely to find work abroad, usually as construction workers or seamen.

Social and economic change in Mulawin

Mulawin's economy, then, is tied closely to global flows of several varieties. In terms of capital, industrial investment at CEPZ provides a substantial proportion of local employment. Flows of people, in the form of overseas migration and inter-regional migration, contribute significantly to the nature of the village's economic and social life. In both of these dimensions, Mulawin appears to be even more deeply embroiled in globalized development than Bunga.

But the data presented in the preceding tables do not give a complete picture of the qualitative differences between Bunga and Mulawin. Even to a naive outsider, the contrast between the two barrios was evident, as the following entry from my field diary suggests:

> Mulawin is a very different place from Bunga, and not just in its lack of agriculture. The roar of the jeepnies passing along the main road, the houses closed off by walls and gates, and the suspicion with which some people initially greet us, all emphasize how much more 'urban' Mulawin is than Bunga. In Bunga, one could pick three or four surnames and their extended clans would represent the entire village; in Mulawin, when I tried to interview a household in the new residential subdivision [Monteverde], their neighbours didn't even know who lived there and three men in the driveway (at least one of whom had a pistol tucked into his pants) were too 'busy' to talk to me.
>
> (Field Diary, July, 1995)

The feelings of longstanding residents also confirmed the sense in which the village had changed. Noting, in particular, the significant in-migration to the village, those who were born in Mulawin spoke of the breakdown of formerly tight social networks and a growing feeling of *anomie* and 'urbanness':

Q: Are there big differences from when you were growing up in Mulawin?
A: Big ones! Of course it's not the same as before, the camaraderie is different. Those who are really from here are different. Everybody was like a relative. Unlike now, when the trend is for those people from other provinces [to come here], it's as if it's every man for himself. Not like before, when if someone was sick you would visit them because they are relatives and friends. It's like Manila now. Manila lifestyle.

(Mulawin resident, 1995; translated from Tagalog)

Others too talked of changing interpersonal relations (*pakikisama*) and a growing unease with the anonymity of their social milieu. Evidently, the tightly woven social fabric that has been the norm in the rural Philippines – and remains largely intact in Bunga, for example – is becoming unravelled in Mulawin. The sheer number of new people in the village means that they cannot be absorbed into existing social networks. This personalized system, through which relationships ranging from personal disputes to business arrangements were structured, has been broken down by the influx of newcomers. Inevitably, tensions and suspicions result:

> When subdivisions were constructed it became disordered, with lots of different kinds of people here. It's difficult to get along with different people, and that's why it's become very difficult since we've had these subdivisions.
> (Mulawin resident, 1995; translated from Tagalog)

> The Visayans are different in their way of thinking. They're not like us Tagalogs. For almost all of them, their attitude when they do something wrong is to think, 'anyway I'll leave this place and they won't see me anymore.' Not like us Tagalogs – our interest is long-term.
> (Mulawin resident, 1995; translated from Tagalog)

The latter quotation comes from a farmer who has surrendered his tenancy rights and lives on business interests established after receiving compensation from the landowner for his eviction. His land currently lies idle and awaits development. The sentiments he expresses are common to many of the long-standing residents in the village who see a way of life, a 'structure of feeling', slipping away.

The survey and interview data presented thus far have provided a picture of the economic and social changes which have occurred in Mulawin due to the village's incorporation into a globalized form of development. The most striking aspect of change in Mulawin is, however, found in the physical landscape, as farmland gives way to residential estates. This process of land conversion represents both a symptom and a cause of these other changes. The rest of this chapter will explore the social, political and environmental context of this process.

The village landscape transformed

The residential areas that have replaced Mulawin's farmland vary considerably both in the quality of housing they provide and in the stage of development they have reached. Many former rice fields just lie idle if the existing landlord or new owners have paid compensation to the sitting tenant farmer to remove him/her from the land. In some cases they are awaiting development approvals, but in most instances they are simply biding their time until the demand for residential land increases still further. Occasionally cattle graze on the grassed fields.

On purely economic grounds, it is clearly irrational to leave the land in this way, but political and legal circumstances dictate otherwise. There are several reasons why it makes sense for owners to remove their tenant farmers. Firstly, if a farmer remains on the land and cultivates rice, the area cannot be zoned by the municipality as non-agricultural, thereby precluding its conversion to other uses. If, however, the farmer is removed and the land is claimed to be unfit for cultivation then legal conversion can proceed. To assist this process, a common tactic has been for owners to destroy irrigation canals and dikes. Secondly, a farmer who still works the land could conceivably claim rights to the land under agrarian reform programmes which legislate the redistribution of land from owners to tenant farmers. In such instances the benefits of land conversion are lost to the owner entirely, and the farmer will be obliged not to sell the land for at least five years. Finally, the longer the tenant is left while land prices increase, the more compensation packages become inflated. This extra compensation would likely far outweigh any rental payments the owner might receive from the tenant if cultivation continued for a few extra years. There are then, economic and legal reasons why landowners are keen to remove their tenant farmers as soon as possible, and these explain the common sight of former rice land sitting idle, occupied only by grazing cattle.

When development of the land occurs, there are several different types of sub-divisions that might be constructed. 'First class' subdivisions include houselots and full service provision – septic tanks for household sewage, concreted roads, pavements, rudimentary basketball courts, and, in one case, even a chapel. In such subdivisions, house lots are sold to buyers who then construct their own homes on the site. At the opposite end of the spectrum is the 'low-cost' subdivision in which one-room, one-storey terraced units are constructed to a standardized design when a purchaser is found. These subdivisions are designed as 'social' housing and most are purchased through contributions to the Government Service Insurance Scheme or national savings schemes such as Pag-Ibig.[18] Such units cost around 160,000 pesos (approximately US$6,000) payable over twenty-five years (although interest charges often double the cost by the end of this period). Whatever the style of subdivision, house lots remain empty until a buyer is found and the new owner decides to construct a house on the lot. The result is often a bleak landscape of occupied houses interspersed with vacant units or empty lots strewn with garbage.

The developers of such subdivisions are mostly of two types. One category consists of wealthy local landowners who have paid off their tenant farmers and have enough capital to construct the basic infrastructure needed for a subdivision, such as roads and drains. The developer would not build any houses, but would instead simply sell off house lots on which others will build homes or keep as speculative investments. In 1991, the Amore and Monteverde subdivisions in Mulawin were selling lots for 350 and 500 pesos per square metre respectively. By 1995, those prices had increased to 1,200 and 2,000 pesos respectively. 'Social' housing, on the other hand, is usually built by large property development companies, with sales offices in Manila. Since restrictions still apply to foreign ownership of land in the Philippines, these companies tend to be predominantly Filipino owned and managed.[19]

The residents of subdivisions in Mulawin are a mixture of young families related

Table 5.10 Residents of Monteverde subdivision, by birthplace, 1995

Place of birth	Number	%
Unknown	12	4.0
Mulawin	87	28.9
Other barrios in Tanza	16	5.3
Other towns in Cavite	33	11.0
Southern Tagalog and Central Luzon	50	16.6
Metro Manila	48	15.9
Mindanao	10	3.3
Northern Luzon	17	5.6
Visayas	28	9.3
Total	301	100

Source: Household survey, Mulawin, 1995

Table 5.11 Residents of Monteverde subdivision by occupation, 1995

Occupational sector	Number	% of working population
Agriculture and resource based	8	5.7
Factory work	83	58.9
Government	9	6.4
Construction and related	14	9.9
Retail/distribution	9	6.4
Transportation and services	18	12.8
Dependency	3	n.a.
Unemployed or at school	157	n.a.
Total	301	100

Source: Household survey, Mulawin, 1995

to longstanding villagers and migrants from outside the barrio. For one subdivision, Monteverde, table 5.10 indicates that approximately two-thirds of the residents are from outside Tanza, and over half are from outside Cavite. This table demonstrates the importance of local factory work in attracting people to the new subdivisions. Over half of Monteverde's working population is employed at local factories, and the CEPZ accounts for the vast majority of this number.

The social and environmental context of land conversion

The chain of events that leads to the construction of such subdivisions starts with the market demand, or speculated demand, for new residential space. A series of legal clearances must be obtained by owners to convert land from agricultural use, as explained earlier, many find ways of circumventing these regulations and others simply ignore them altogether. The start of the process for farmers comes when the landowner approaches with the suggestion that they might wish to sell the land, and tries to evaluate the tenant's likely demands for compensation:

> What they will do is they will approach us. Naturally we first talk about farming. [Then they say] 'If perhaps I were to sell this land, would you be agreeable to my suggestion'. And the tenant says, 'If you can grant our rights, we could reach an agreement'. Two or three times they'll approach you. Of course, he is the one who is more eager to sell. In our case, since what we want is to avoid a problem, then we agree. That is how the system works.
>
> (Mulawin resident, 1995; translated from Tagalog)

Negotiations inevitably vary according to the individuals involved, but the social relationship between landlord and tenant, going beyond their economic arrangements, means that farmers often feel 'ashamed' or 'embarrassed' to negotiate as strongly as they might. Farmers feel unable to go beyond certain culturally prescribed bounds:

> It is inappropriate for you to act superior to the owner of the land.
>
> (Mulawin resident, 1995; translated from Tagalog)

> For us, we just go along with the agreement, because it's theirs, and it's inappropriate for us to say we don't want to. It will appear that we are becoming greedy over it.
>
> (Mulawin resident, 1995; translated from Tagalog)

The settlement that is eventually reached between landlord and tenant will usually provide both a cash payment and a small parcel of land on which the tenants can build houses for themselves and for their children. Cash payments have escalated in recent years but individual settlements vary according to the location of the land involved and the negotiating skills of the tenant. In 1995, a typical compensation package amounted to approximately 500,000 pesos for each hectare of farmland (or 50 pesos per square metre) and a house lot of 1,000–2,000 square metres. The selling price of land to a developer, meanwhile, might be many times greater. One price being quoted by a landowner in Mulawin in 1995 was 350 pesos per square metre (3.5 million pesos per hectare).

Various legal structures, described earlier, regulate the conversion of agricultural land to other uses. But like elsewhere in Cavite, many of these regulations have been circumvented if not technically breached in Mulawin. Despite clear legal regulations against converting land that is either eligible for redistribution under agrarian reform or is irrigated, land that falls into both categories has been converted, or has been taken out of production, in Mulawin. The reasons for this can be identified in the social and environmental context of the land market and rice cultivation. Six contextual factors can be identified in Mulawin.

Firstly, farmers themselves are usually poorly informed of their legal rights. Cases exist, for example, of farmers who are *de facto* tenants on a piece of land, but are not legally registered as official tenants. In other instances, registered tenants with Certificates of Land Transfer under agrarian reform programmes have seen their transfers cancelled for no official reason (although usually due to the intervention of the landowner with agrarian reform officials). In all such cases, a system of verbal agreements and unwritten understandings conflicts with a legal system of documentation and regulation. In such circumstances, it is invariably the educated landlord, with high social status, sufficient resources to bribe officials, and access to legal counsel, whom the situation will favour. But it should also be added that many potential agrarian reform beneficiaries would rather enjoy a lucrative cash settlement than continue farming with the added burden of amortization payments to the Land Bank. Thus all land is effectively negotiable if the two parties can reach an agreement. In one case in Mulawin, a farmer sold his rights to two of the three hectares he farmed and received a cash settlement and the remaining hectare under agrarian reform. He was, legally, entitled to all three hectares under agrarian reform but could not then have converted the land for a minimum of five years. In another case, a farmer said that he had 'surrendered' his right to agrarian reform, even though legally there is no provision for doing so. But, as a *barangay* official who was also present at the interview noted, 'with every law there is an exception, if two persons agree with each other'.

Secondly, a landowner's ability to circumvent regulations in this way is a product of more than just access to political and judicial systems. As noted earlier, a farmer's association with the owner is more than just a legalistic, landlord–tenant relationship. The bond between the families may date back several generations, and the landlord might, for example, be a godparent to the tenant's children. Consequently, tenants are reluctant to try forcing their legal rights and souring personal relationships:

> That's it, that's their proposal. Of course, you're ashamed because we've been together for a long time. We don't want them to say we were greedy when the [agrarian reform] law came, doing everything by the letter of the law. We don't want that, so I just accepted it. Even though what is happening is painful, there's nothing we can do about the situation.
>
> (Mulawin resident, 1995; translated from Tagalog)

A third context in Mulawin which effectively removes any remaining obstacles to land conversion is the compliant political environment in which developers find themselves. The key regulatory role that municipal officials perform is in ensuring that land proposed for development falls within the area zoned for that purpose and does not merit protection under any of the provisions described earlier. When asked why irrigated land has been converted even though the law states that it is protected, one *barangay* official simply replied: 'perhaps not, because with our current system [of government] it passes through' (translated from Tagalog). Another government official noted the vested interest which a municipality has in approving land conversion, which often includes skimming off a percentage of the sale price of the land. In some cases, however, developers also ensure political goodwill by making personal payments or gifts of houselots to local officials. One farmer in Bunga commented: 'I don't know why [it happens], but I really don't like it. We cannot do anything about it. For example, it's our leaders who let it happen. They [developers] pay them so that it will be built. There's nothing that can be done' (1995; translated from Tagalog). This was a practice that was evident not just in the hearsay of villagers, but also in the experiences of land developers whom I interviewed. Thus, where political officials, landowners and farmers themselves all have substantial vested interests in seeing the land converted to non-agricultural uses, there are no parties left with a legal interest in the transaction to protest. And, as one Mulawin resident noted, 'even if you protest, the person to whom you take your grievance will have something to do with the project'.

The local politics of individual enrichment and bureaucratic corruption cannot, however, be divorced from the wider politics of development that lie behind the changes occurring in Cavite. In broad terms this refers to a political agenda based on a globalized form of development. But more practically this strategy translates into the priorities set for government agencies. In the case of irrigation authorities, for example, some farmers complain that the irrigation system, while constructed with substantial loans from the World Bank for the Second Laguna de Bay Irrigation Project, is in fact neglected on the ground, with dike maintenance and water supply inadequate for their needs: 'It's because of the interest of many government officials. That is a very big question for us.' Many also question the priorities of the national government concerning food security, as so much rice-producing land is converted. Some believe that the kickbacks involved in rice importation mean that officials actually favour food imports over local production. According to this body of opinion, not even the National Food Administration has the best interests of farmers at heart.

There are, therefore, two forces working against the proper functioning of these government services. One is the priorities set by the administration in terms of development strategies, and particularly the relative importance given to agricultural versus industrial development. Thus appeals against land conversion fall on deaf ears at both provincial and national scales. The other is the susceptibility of regulatory frameworks to influence through bribes or personal favours. One land developer talked frankly about using high-level government contacts in Manila to secure a land conversion clearance from the Department of Agrarian Reform.

A fourth context for land conversion is provided by the types of socio-cultural changes described earlier and in chapter 4 (see also Kelly, 1999c). A shift to urban, industrial and overseas work as the primary sources of livelihood in Mulawin has rendered farming an anachronism in the minds of many of the younger generation. As other opportunities have presented themselves, young people have shifted their aspirations away from a rural life and towards other forms of work. This provides an unavoidable part of the calculus for tenant farmers confronted with offers of cash compensation to remove them from their land. Several farmers expressed the sense of low self-esteem to which their profession had sunk: 'they belittle themselves, that's how they feel about themselves, they pity themselves when they are farming' (Farmer in Mulawin, 1995; translated from Tagalog). Another farmer in Mulawin noted the ways in which peer pressure can reinforce such a 'modern' attitude to work: 'Modern times is a very strong influence, For example, your son is doing something, like hoeing, and a group will pass by, and they'll call to him . . . "that's not the real life" ' (Farmer in Mulawin, 1995; parts translated from Tagalog). Some related these changing attitudes directly to the influence of the media. One farmer, ironically, delivered such a comment in hybridized 'Taglish':

> Because the new generation and the TV program, they show about the different standard of living of some people. I mean, they never study the effect of us copying things from other countries shown on the TV programs. Young people are thinking that this is the better way of life . . . So that's why some people dream of the new lifestyles . . . So they hate farming.
> (Farmer in Mulawin, 1995; parts translated from Tagalog)

This broader disinterest in farming translates into a sense of futility among farmers if their own children are disinclined to inherit the tenancy after them. In this context, where farmers see little prospect of passing their tenancies onto their children and where they are being offered substantial compensation packages to stop farming, acceptance seems the only choice. At least with substantial capital and usually a houselot on a new residential subdivision they will be able to ensure the education of their children and perhaps provide them with space to build a house for their families.

A fifth factor that facilitates the process of land conversion is the set of environmental conflicts that ensue when agricultural and non-agricultural land uses are juxtaposed. Just as labour market changes have put farming in Bunga under considerable strain, so the physical transformation of Mulawin's landscape presents difficulties for residual farmlands. Even before the construction of a new housing estate begins, problems can start for farmers adjacent to abandoned fields. If, as is commonly the case, the idle land is grassed-over and grazed by untethered cattle, farmers find their crops being eaten or trampled by straying animals. Farmers may also find their fields waterlogged in the rainy season if the dikes (*pilapil*) on adjacent land have not been maintained, thereby destroying the water management system and causing excess runoff. Once construction begins, vegetation and any remaining *pilapil* are removed and the problems of excess runoff during the rainy season are exacerbated. For one farmer, whose land borders on a subdivision currently under construction, rain and soil washing onto his land have severely reduced yields: 'when it rains, their soil mixes with water and engulfs my crops. It's like *lahar*' (Mulawin resident, 1995; translated from Tagalog).[20] The decline in yield for the farmer in question has been from 90–5 *cavans* per hectare in 1992, to 35–45 *cavans* in 1995.

Plate 10 Refuse from a new residential subdivision clogging an irrigation channel

Even after construction, farmers face water management problems during the rainy season as runoff is further increased by the concrete roads and drainage systems that, in many cases, flows directly into irrigation channels for adjacent fields. In another case, subdivision and house construction on a social housing project left a farmer unable to plant any crops in a five-metre strip adjacent to the development. During the completion of the houses, workers had discarded glass and steel fragments onto the field rendering it too dangerous for rice cultivation, which involves bare-footed wading in the soft mud of the paddy field.

After the completion of subdivisions, neighbouring farmers face further conflicts with residential land use: 'The effect [of the subdivisions] is to pollute the surroundings. For example, the housing close to that field. All the waste from that place will go into the field. It will be polluted' (Mulawin resident, 1995; translated from Tagalog). Solid waste disposal and household sewage are particularly problematic for nearby farmers. Septic tanks, which are prone to leakage, rather than centralized sewage treatment, are used in most subdivisions, and when leakages or overflow occurs, the waste drains directly into local irrigation channels (see plate 10). A more immediate problem, however, is the system for solid waste disposal. Garbage removal occurs once a week, but residents often leave their garbage bags beside the main road or simply dispose of them in nearby irrigation or drainage canals. This has caused problems of flooding during heavy rain storms in Mulawin and has restricted water flows to local farms.

As the issues of drainage and waste disposal suggest, a major source of conflict between urban and agricultural land uses is water management, and particularly the supply of irrigation water. Some farmers believe that irrigation systems have been deliberately neglected by those who see financial opportunities in land conversion, but water supply also suffers from the environmental externalities of conversion. Irrigation canals clogged with garbage are the most visible indications of these externalities, but in some cases the construction of subdivisions has completely cut off irrigation supplies to particular pieces of land. Problems also arise during the construction process. The change to the water management system and the erosion of top soil into irrigation channels leads to the gradual accumulation of sediment. This is particularly acute at the furthest reaches of the irrigation system where water comes to a stop and deposits sediment. In these areas the shallowing of the channel and the reduced gradient mean that it is almost impossible to draw irrigation water. Even in July of 1995, when the rains had already started, large areas of residual farmland in the western portion of Mulawin were left unplanted as farmers waited for irrigation water to be supplied. Many farmers ascribe the neglect of the irrigation system in this way to a lack of political motivation and vested interests in seeing land converted.

A sixth factor behind land conversion is the fact that, in addition to all of the problems described above, farmers in Mulawin must also cope with the same labour shortages as their counterparts in Bunga. Remarkably, given the village's exclusively agricultural landscape just a few decades ago, only one household in Mulawin relies on agricultural labour as its primary source of livelihood, meaning that tenant farmers must seek out harvesting teams from further afield, particularly those visiting from Batangas. Rice planters are also difficult to recruit locally, leaving most farmers to seek assistance in neighbouring barrios:

> If it were not for the problems in harvesting and planting, and if we did not succumb to what is being offered to us, then the subdivisions could not be built. We often think that the time will come when you can no longer plant or harvest. In the past, the problem of farmers was just water; harvesters and planters were plentiful. It is ironic for the farmer now – while the farmland is getting smaller, and the population is getting larger, why is there such a shortage? Now there's no one to harvest, no one to plant.
> (Mulawin resident, 1995; translated from Tagalog)

For minor tasks other than harvesting and planting, most farmers in Mulawin rely on a pool of migrant labourers, many from the Visayas, living in small huts in the rice fields. These men work for the farmer, whose land they live upon, when called to do so and are paid only for the days they work. At other times these workers, many of whom have come to live in Mulawin permanently, will harvest for other farmers, weed crops, cut grass, mend dikes and perhaps plant vegetables on borrowed land on their own account. It is these workers, at the economic, social and geographical margins of the village that have the most to lose in the land conversion process, being deprived of their source of livelihood, but receiving no compensation in the land conversion transaction.

Six factors, therefore – farmers' legal ignorance, the social relationships between landlords and tenants, the context of political compliance, sociocultural change, environmental conflict and labour shortages – all contribute to a situation in which the legalities of land conversion can be side-stepped through mutual agreement with a tenant farmer. These social, political and environmental relations, locally embedded but with connections at other scales, create a context in which land conversion driven by globalized development can rapidly proceed. They thus form the multi-scalar local context in which globalization is constituted.

Farming on the urban fringe

Despite the abandonment of agriculture across large areas of Mulawin, some farmers still attempt to cultivate crops, particularly in areas set back from the national road. My household survey, combined with interviews, revealed twelve households for whom farming as tenants represented their primary income. The land area they farmed totalled approximately thirty-two hectares. Despite the variety of pressures described above, in 1995 these farmers were still attempting to derive a livelihood from agricultural activities and showed signs of intensification and diversification into new crops. Successful operations in pig-raising, freshwater aquaculture and melon cultivation, for example, all suggested that with adequate inputs of capital, labour and technical knowledge, the agricultural sector can be viable in an urbanizing/industrializing environment.

One farmer's experiment with aquaculture has proved to be particularly successful. Together with a financial backer from another barrio in Tanza, the farmer has converted approximately one third of a hectare from rice land to ten freshwater fish ponds. The start-up costs for the enterprise were substantial: 50,000 pesos for manual labour to excavate the ponds; 30,000 pesos to construct a deep well and water pump to supply the ponds with fresh water; 50,000 pesos to stock the ponds with fingerlings of African Catfish; and around 200,000 pesos to provide daily feed to each of the ten ponds over the course of one harvest cycle lasting around three months. The returns on this investment are, however, impressive. Excluding the capital costs involved in establishing the operation, each pond generates a profit of 10,000–20,000 pesos at each harvest, with up to three harvests per year. Thus over a single year, the fishpond operation can generate an income of approximately 500,000 pesos. Furthermore, freshwater fish farming provides a complementary land use for rice cultivation, as the water drained from the pond after a harvest is rich in digestive by-products which fertilize the agricultural land.

The lack of sufficient capital, an appropriate site and technical knowledge all prevent other farmers from embarking on a similar initiative, but many have, nevertheless, intensified their production and diversified the range of crops grown. In particular, the cultivation of muskmelons has become an important part of the land use cycle for farmers in Mulawin since the mid-1980s. During the dry season, when there is seldom enough water to plant rice, farmers devote a large part of their land to the crop. An individual farmer might be capable of cultivating around 3,000 melon plants (or approximately 0.5 hectares), but for a larger area extra help is needed. Hence farmers enter into a variety of arrangements with others to cultivate melons. In some cases the farmer will simply employ helpers to assist on his land in the melon cropping, in others the farmer will act as a financier and share the harvest with someone who will provide the labour input in caring for a patch of melons. Alternatively farmers may simply lease, or even lend without charge, a part of their land to someone else for a period of three to five months. In 1995, the rental charge in such an arrangement ranged between 2,500 and 4,000 pesos per hectare, depending on the distance of the land from the road and the relationship between the two parties. Particularly active in this enterprise are fifteen to twenty farmers who have sold their own tenancy rights to property developers and who are using their compensation money as capital to establish melon-planting operations on land sub-rented to them by remaining tenant farmers.

The returns from melon cultivation are substantial, yielding up to 100,000 pesos per hectare in gross income before labour and input expenses are deducted. The net income will depend on the amount of family labour that is available, but since unlike rice, melon cultivation does not require major episodes of intensive labour usage at planting and harvesting time, the expenses are lower. A farmer might make around 70,000 pesos from one hectare of melons, and two such croppings are possible in the dry season between October and April. Furthermore, since rent to the landlord is paid in the form of *cavans* of rice from the rainy season harvest, the income from a dry season melon crop is free from any deduction. Melon cultivation, then, provides a stark contrast to the economics of rice cultivation described in chapter 4. Limitations to its spread are provided only by the need for initial capital, sufficient labour inputs, occasionally water shortages, and the need for specialized knowledge of cultivation techniques. The accumulation of technical knowledge over the last ten years in Mulawin has made it, and Tanza generally, a major centre for melon cultivation in Cavite. This has in turn brought buyers from the wholesale market at Divisoria in Manila to the area which has made the marketing of the crop easier.[21] The only further limitation on cultivation is the rapid conversion of the land base to other uses.

Conclusion

As the example of melon cultivation demonstrates, farming *can* be profitable in Mulawin, and by extension in this zone of globalized development more generally. The marketing advantages of being close to Manila mean that orthodox predictions of agricultural intensification around a large city *could* hold true. But globalized development is driving a process of urban expansion involving the widespread conversion of agricultural lands in a process closely related to the social, political, and environmental context of villages like Mulawin. The result is that instead of co-existing profitably with globalized development, agriculture is being effectively 'squeezed out' (see Kelly, 1999a).

Social relations between landlords and tenants mean that those working on the land, and potentially profiting from cultivating crops such as melons, are not the actors who principally decide upon the fate of the land as a commodity. This tenancy structure combines with a relationship of personal power and influence in which farmers are unable or unwilling to assert their legal rights to continue working the land. Political structures in the form of pervasive bureaucratic corruption create a context in which a powerful logic of rent-seeking capitalism can operate freely. Finally, environmental conflicts between agriculture and urban land uses create tensions that reduce the productivity of rice farming.

But these locally embedded social relationships are also connected to policy priorities set at higher levels, and in other scales, of politics. In a context where development strategies favour industrial expansion over support to the agricultural sector, it becomes easier for the needs of farmers – irrigation, inputs, marketing etc. – to be neglected. In many ways, the bureaucratic corruption that also inserts itself into these institutions is less a cause than a symptom of the low priority they are given. There is, then, a complex web of relationships that link the broader discourses of globalization and the priorities it sets to the local processes of landscape transformation in which it is played out. But what is clear from the evidence in this chapter is that the process of globalization, as it is imprinted on the physical landscape of a village such as Mulawin, can only be understood in the context of political, social, economic, and environmental conditions at scales much smaller than the global.

6

RESISTING AND REIMAGINING THE GLOBAL

The last three chapters have examined the mediated process of globalization as it is experienced in particular lives and localities in the Philippines. These changes have not, however, been received passively. In this chapter I will consider some of the reactions to globalization in several different spheres of political action. At the national level there exists a strong line of leftist nationalist thought that opposes globalizing trends on the grounds that they create a situation of neocolonial dependency and exploitation. This school of thought is articulated by such leftist academic and journalistic writers as Renato Constantino, and such nationalist political figures as Wigberto Tañada. It must be seen not just as a brand of generic socialist ideology, but also as a local reaction to the distinctive colonial and post-colonial history of the Philippines described in chapter 2 – a reaction that draws some of its vocabulary from the lexicons of anti-imperialism and marxism. There is also, however, a component of mainstream political thought in the Philippines that appeals to economic nationalism – one that has found increasing appeal with the vulnerabilities exposed by the financial crisis of the late 1990s.

A second stream of anti-globalization thought and action is to be found in the praxis of Filipino civil society associations, usually referred to as non-governmental organizations (NGOs) and people's organizations (POs). These organizations range from radical leftist guerrilla movements to church-based community groups. They also include issue-based social movements and broader umbrella organizations. Together they form a lively and kaleidoscopic progressive movement in the Philippines that in many ways represents the *de facto* political opposition in a context where formal political positions are dominated by competing personalities rather than contending ideologies.

The third source of resistance to globalization is to be found in *ad hoc* episodes of everyday resistance, some of which rise up into transient issue-based and localized oppositional movements. Usually, these movements are responses to the impact of globalized development on individual rights and livelihoods rather than a broader and coherent anti-global philosophy. In many instances, people are powerless to resist the impacts of globalized development, but in a few cases notable successes have been achieved. I will describe two such cases – one an aborted power plant project, the other Cavite's provincial election of 1995 – in which local power structures were successfully challenged.

Although this chapter considers each of these sources of anti-globalization rhetoric and activism separately, in fact there exists considerable overlap between them. The ideas of leftist nationalists frequently form the ideological foundations for the positions taken by progressive groups, and in fact the personalities involved in both intellectual/political and activist spheres are frequently the same. Equally, the everyday political resistance of issue-based and localized struggles is frequently facilitated by the involvement of larger NGOs or POs.

Leftist nationalism: globalization as neoimperialism

Much of the inspiration for anti-globalist positions has come from such nationalist intellectuals as Alejandro Lichauco and Renato Constantino, who have authored numerous tracts on Philippine political economy, identity and history, mostly inspired by the marxian *dependencia* movement in development theory.[1] Constantino, in particular, stands out, with writings that span the last fifty years and continue today through syndicated newspaper columns available on the internet. He argues that the economic plight of the Philippines is deeply bound up with political-economic ties and cultural identity rooted in colonial power relations. He traces current thinking towards economic, political and cultural relations with the rest of the world to the political, religious and educational institutions established by Spanish and American colonialism to foster dependent development and a colonial consciousness (see chapter 2).

The central purpose of Constantino's writing is the development of a counter-consciousness in the service of cultural decolonization. His approach is to encourage mass education in the historical legacy of colonialism and thereby to encourage a critical approach to Philippine relations with the outside world.[2] In this way, Constantino has tried to articulate a global imaginary in which the Philippines is not subordinated to the demands of either imperialist powers or global capital. In a booklet called *A Filipino Vision of Development*, for example, Constantino argues for a series of measures to carefully regulate the relationship between the Philippines and foreign capital in the process of industrialization (Constantino, 1991). Firstly, instead of attempting to attract as much foreign investment, of whatever type, as possible, Constantino argues for a planned and regulated investment programme for industrialization in diverse sectors and an emphasis on local sources of capital. Secondly, Constantino contradicts the prevailing wisdom that protectionism is somehow regressive and arcane and argues that locally owned enterprises should indeed be protected from outside competition. Thirdly, rather than basing development on export manufacturing using cheap labour, Constantino insists that governments should foster strong domestic demand through higher wages and redistributive reform, particularly in the agricultural sector. Priority should also be given to national security in essential sectors such as food production. Finally, Constantino criticizes the 'race to the bottom' mentality that the global economy fosters in developing countries and deplores the competition that forces countries to compete with ever more enticing tax incentives, cheaper labour, and ownership privileges for global capital.[3]

Constantino has also addressed the discourse of globalization directly. In a column entitled 'Globalization hype', he writes:

> The Philippine government is faithfully following the globalization prescriptions of international capital, seeing these as the spurs needed to push the country out of the economic doldrums, and make it forge ahead in the GNP race not too far behind its neighbors. . . . The now familiar name of the game, 'global competitiveness', is pushing nations to prove themselves in the economic arena by producing 'export quality' products at the cheapest possible price in order to fare well in the ultimate testing ground – the international marketplace. . . . Globalization is a disease which is now ravaging the lives of the more vulnerable peoples of the world.[4]

Constantino's arguments can be criticized on several grounds. Firstly, some historians have objected to his partisan and overtly political scholarship (May, 1987). Secondly, Constantino adheres closely to a class-based analysis that assigns the primary historical struggle to the 'working class' while neglecting other axes of resistance based on gender, ethnic or religious identities in the Philippines. Such a framework also tends to neglect the realities of political action in the Philippines based in complex structures of factionalism, clientalist networks and bossism, which have been the subject of vigorous debate in recent years (Sidel, 1998). Finally, Constantino's arguments about globalization tend to draw upon an understanding of power relations in the global economic system which often neglects a critical analysis of how globalization is embedded in local social relations. Thus, Constantino neglects precisely the power relations at a smaller scales described in earlier chapters of this book.

Constantino does, however, articulate an alternative imagination of the Philippines' place in the world that is widely shared by others in the progressive movement to be discussed later. The result is the creation of an intellectual space for resistance to the dominant view that represents globalization as an inevitable and irresistible destiny.

While Constantino's works may represent a rather extreme equation of globalization with neoimperialism, the type of economic nationalism that he espouses finds echoes in political voices closer to the mainstream of Philippine political debate. Some are rooted in the economic nationalism of the 1950s and early 1960s, articulated by such individuals as Jose Diokno, Lorenzo Tañada and Claro Recto. Such establishment figures advocated policies of mild protectionism, the withdrawal of US forces (until 1991, this meant the closure of the US bases) and economic/political influence, and a 'Filipino First' regulatory framework. Their nationalist sentiments were developed in a period of continued US economic and political dominance, for example under the Bell Trade Act of 1946 that ensured parity rights for US citizens in terms of access to the Philippine economy (Cullather, 1994).

An extensive literature emanating from economic nationalist thought circulates in the Philippines through such organizations as the Foundation for Nationalist Studies and Karrel Press Incorporated (see Arcellana, 1996; Constantino, 1989). The flavour of this body of thinking is captured in an excerpt from a speech by Recto:

> I need not stress the point that when I speak of industrialization, I mean nationalist industrialization, that is, the industrialization of our economy, the Filipino economy, not merely the industrialization of the Philippines in a territorial sense. . . . [F]oreign direct investments, as distinguished from foreign loans, not only will channel the nation's wealth into foreign hands, but will fail to promote the industrialization of the Filipino economy because it will not help in the formation of Filipino capital . . .
> (Claro Recto, Speech entitled 'Industrialization, the only cure for our economic and social ills', 24 June 1955; quoted by Constantino, 1985)

Such thinking continues to enjoy broad appeal in the Philippines and works by Recto, Constantino and others are widely circulated. Economic nationalism has also continued to find a voice from within the contemporary political establishment. Senator Wigberto Tañada, for example, in commenting on the incentives for foreign investors in the Medium-Term Philippine Development Plan 1993–8, asked:

> Are the economic planners afraid to be called pro-Filipino? Are they afraid to be called nationalistic? Is there no room for nationalism in our development plan? Are we producing a development plan for foreign investors alone? Are we not supposed to craft a development plan that will really empower the people, meaning that the great masses of our people will not only become the prime beneficiaries of the plan but will also become the leading actors in the investment and development process? If not, we will be a historical oddity. For here are strange people who have lost their sense of nationality and nationalism that in the process they have become utterly dependent on foreign advice, foreign investments, and foreign assistance.
> (Tañada, 1993: 94)

More recently, this strand of economic nationalism has found new voices as foreign direct investments have flowed into the country in the 1990s and as the crisis of the latter part of this decade has exposed the vulnerabilities of such a development strategy. Even mainstream political figures such as Manuel Villar, Speaker of the House of Representatives, have urged renewed consideration of development strategies based on globalization in the light of the regional economic downturn. In a speech on 5 October 1998, he argued that 'unless appropriate measures are made by both the individual countries themselves and the international community, the asymmetries implicit in the process of globalization are bound to intensify uneven development . . . We must tame the forces of globalization to service only development and prosperity.'[5] President Estrada too has moved governmental rhetoric closer to a nationalist line, placing emphasis on agricultural development, for example, but at the same time attempting to pursue foreign direct investment and maintain the liberal regulatory regime created by his predecessor.

Filipino progressive movements: BINGOs, PONGOs, FUNDANGOs

Much has been made of the emergence of civil society in Asia in recent years. For some, the autonomous groups that have emerged between the state and the individual with the explicit purpose of fostering social change represent the most promising arena for the expansion of democratic space in Southeast Asia. In Singapore and Thailand, for example, despite the diverse contexts, commentators have suggested that effective political alternatives are likely to emerge not from formal parliamentary democracies but from the resistance, persuasion and dynamism of citizen's groups that present alternative ideas and approaches (Chua, 1995; Quigley, 1995; Rodan, 1996, 1997; Tay, 1998). In the Philippines too, where civil society movements have perhaps reached their highest level of development in the region (particularly since the overthrow of Marcos), this alternative forum for political expression and action has been increasingly important.[6]

While some would apply cultural arguments – the 'freedom-loving' Filipino, the 'Latin' temperament – to explain the relatively dynamic state of Philippine civil society, it is likely that contextual and historical factors provide more plausible explanations. The gradualist and non-violent nature of Philippine independence in the 1940s meant that the leftist and nationalist resistance movements were left only partially satisfied and so continued to press for nationalist economic policies of the type described above. Subsequently, the declining legitimacy of the Marcos regime increased the numerical strength and political power of leftist movements, including armed insurgencies such as the New Peoples' Army. Thus, after the downfall of Marcos, which many civil society groups helped to bring about in the so-called 'People Power' revolution of 1986, there was a mushrooming of generally leftist groups of intellectuals, activists, peasants, environmentalists and unionists, ranging from the established Catholic church, which had engaged in 'critical collaboration' with the government, to the armed revolutionaries of the NPA. In between emerged a plethora of issue-based groups, umbrella organizations and ideological movements focusing on issues such as gender relations, peasant rights, land reform, national debt, worker protection, environmental protection, overseas workers, native peoples, electoral processes, urban housing and squatters' rights, livelihood co-operatives and public health. With Congress still largely dominated by elite interests it has been these groups that have articulated alternative political visions and policy directions. The proliferation of acronyms give some indication of the diverse positions taken by these groups in relation to the political status quo – BINGOs (business-initiated NGOs); GRINGOs (government-initiated NGOs); COMENGOs (the fly-by-night variety); FUNDANGOs (funding agency dependent or initiated NGOs); PONGOs (people-oriented NGOs) (Tigno, 1997). Just as their relationship with government and business interests vary, so too do the ideologies and beliefs of such groups differ. Even within the generally leftist progressive movement there have been deep divisions in the 1990s, reflecting a split based on ideological and personal differences within the Communist Party of the Philippines and National Democratic Front.[7]

Many of these groups have articulated positions on the issue of globalization. At one end of the spectrum are the ideologues of the radical Maoist left who castigated the 'US–Marcos dictatorship' and hyphenated the Aquino and Ramos regimes in a similar way, while demanding the removal of foreign capital from the national territory (Goodno, 1991). Less extreme are those groups focusing on specific manifestations of globalization such as the plight of overseas contract workers, women involved in the sex trade, farmers impoverished by global commodity chains, and workers in foreign-owned factories.

One such organization, the Kilusang Mayo Uno (May First Movement), an association of trade unions, has outlined the following vision of 'nationalist industrialization':

> the country is trapped in a vicious cycle of underdevelopment where we are forced to export all we can, import in order to survive, and meanwhile remaining backward and non-industrialized basically.
>
> This situation, the KMU believes, is the root cause of unemployment and underemployment in the country. The task of the government, therefore, is not only to come up with short-term, patch up solutions to the problem but to pursue a long-term program for genuine industrialization that will essentially restructure the Philippine economy. By this, we mean the expansion of local industry, with the domestic market, not the world market, as its base. This is what the KMU basically means when we call for nationalist industrialization.
>
> ('KMU Position on Job Creation', n.d., reprinted in Schirmer and Shalom, 1987: 376)

More recently, the explicit focus on globalization in the Ramos administration's development policy has provoked attacks and campaigns by other civil society organizations. The Ibon Foundation, for example, an independent policy advocacy and research organization based in Manila, has criticized the liberalization policies undertaken by the Ramos government and embodied in foreign investment legislation – most recently the proposed Multilateral Agreement on Investments (MAI) promoted by the industrialized nations through the Organization for Economic Cooperation and Development. In a special issue of its 'People's Policy and Advocacy Studies' newsletter, Ibon argued that:

> The advent of globalization signalled the increasing subjugation of human economic, political, social and cultural rights to that of corporate interests. Liberalization has legitimized increasing exploitation of workers and natural resources, and caused greater poverty to peoples of the Third World.

> The Philippines is already experiencing the perils of liberalization. Even as the government proclaims sound economic fundamentals, the real situation of the Filipino masses, faced with unemployment and sinking living standards, belie the claims. Therefore, further liberalization, as what the MAI would do, is contrary to the interest of the Filipino people.
>
> (Ibon, 1998: 15)

Other groups have brought together scholars and activists and have sought to establish international linkages. In February 1998, for example, the Kilusang Magbubukid ng Pilipinas (Peasant Movement of the Philippines or KMP) acted as the Filipino co-convener of a conference in Geneva for 'The Peoples' Global Action Against Free Trade and the World Trade Organization (WTO)'. The conference aimed to 'lay down the basis of an alliance in the form of a manifesto . . . and discuss specific courses of action, including the coordination of decentralized protest actions all over the world parallel to the Second Ministerial Conference of the WTO scheduled on May 18–20, 1998'.[8]

The draft manifesto of the conference made the following statements concerning globalization:

> Capitalism inevitably matures into imperialism. This has been the cause of the two world wars. Capitalism must globalize itself through political and economic machinations and in the process brings out fatal flaws inherent in it.
>
> In its current sense 'Globalization' means further rearranging the international economic order to avert the crisis of capitalism. It means the further dismantling of barriers to the free movement of capital to seek maximum profits even as the great industrial powers exert all efforts to protect their own saturated markets.
>
> The consequences to the peoples of the North and South are disastrous; falling wages, cut in social services, lack of job security, displacement of peasants, etc. In short, the dominance of international capital on the economies of both the North and South is tightened, further closing the avenue to a development of self-reliant economies.[9]

The statement is clearly inspired by a Marxist–Leninist understanding of capitalism and neoimperialism, but perhaps more importantly the manifesto and the conference at which it was discussed highlight not just the globalization of economic processes, but also the globalization of resistance movements to these processes. The KMP was, in this case, collaborating with more than ten other organizations from around the world.[10] Thus while globalization is being opposed by the KMP and other organizations, it can also be interpreted as a set of processes through which their resistance is facilitated and strengthened.

Discussions regarding the desirability of globalized development have been given a new sense of urgency by the financial crisis that spread across East and Southeast Asia in 1997–8. Thus a conference organized by the BAYAN New Patriotic Alliance and the Ibon Foundation in November 1998 in the Philippines took as its theme 'Alternatives to Globalization' arguing that 'the people the world over are now seriously asking for discussion on alternative strategies and paradigms to neo-liberal globalization'. They go on to suggest that 'we need to develop a critique of the monster that is neo-liberal globalization'.[11]

Several points are worth noting in the rhetoric employed by these organizations. Firstly, they recognize the need to seek out alternatives to neoliberal globalization, but just as the political orthodoxy represents globalization as inevitable and irresistible, so the NGOs that oppose it tend to recreate this image through references to a 'monster' that is the product of the inherent logics of capitalism. The monster is clearly super-human, and the only solution that is left open in such an analysis is a reconstitution of economic and social life in the form of some alternative to capitalism – a solution that leaves limited space for manoeuvre. Secondly, as noted above, a key element of the strategy used by NGOs has been the formation of international alliances with like-minded groups around the world. If globalization is seen as more than simply the intensification and expansion of economic linkages through trade and investment, then clearly there is a progressive dimension as well. Globalizing channels of communication and interaction are precisely what allow this process of alliance formation among civil society groups to occur.

Civil society organizations have not, however, limited their activities to the production of alternative imaginations of national development. In addition to reproducing varied forms of leftist nationalism, they have also engaged in active resistance to the consequences of globalized development. This has usually taken the form of involvement in, and organization of, transient protest movements against particular developments. By linking with localized and issue-based social movements in this way these groups have tried to synthesize local scale oppositions to exploitative social practices associated with globalized development into an alternative vision of what development should mean. In the next section we will examine an example of this type of praxis in more detail.

Everyday resistance and social movements

Many of the oppositional sentiments articulated by leftist-nationalist intellectuals and non-governmental organizations are echoed in opinions expressed by villagers in Cavite. One farmer identified the need for nationalist industrialization in this way:

> We don't need to look to other countries. It's because our management here are selfish, they don't want to share. Investors are coming here, aren't they? Couldn't our own people do the same thing? Like the Ayalas, they're very rich. They could afford to build ten factories. They could do that. And they're really sharp people. But they still want other countries, and they

come in here for their own benefit.... Whatever benefits they gain would go to us if only our own people were the ones helping us. But no ... I heard they're even going to Australia to get foreigners to come and develop our country.

>(Farmer in Bunga, 1995; translated from Tagalog)

Aside from regret at the shortage of labour created by rapid industrialization, the principal issue over which individual opposition to globalized development arises in Cavite is the land conversion described in chapter 5. It is undoubtedly the case that many farmers would prefer not to see their land, and the land around them, converted to other uses. Many farmers to whom I spoke expressed an emotional attachment to the land and disappointment, frustration and anger that the conversion was proceeding so prolifically. Two lengthy but eloquent quotations capture these sentiments:

> That's really a rice field, [but] they built a subdivision. They will cause great hunger among the people. They have no pity, the landowners have no pity in destroying rice fields. These have been rice fields for a long time. Those rice fields have existed since Spanish times. Even before the Spanish times they were rice fields. Now, I think people have become greedy. I don't know what it all means. If they are constructing subdivisions, we have large areas of uplands where they should build them.
>
> If we really lack a place to build houses here [in Cavite], it should be done there in the uplands. We should do it there, not here in the rice fields. Can you imagine, these are rice fields where we have harvested, where we have grown lots of food. Now what will we eat? There's certainly not enough for us now. We will suffer a shortage.... We were pushed to the edge. Is it reasonable for them to sell it? They knew we are working on it. They should not encourage us to sell, because that's where we get food ... We should love it forever because that's where we get our existence.
>
>> (Farmer in Mulawin, 1995; translated from Tagalog)
>
> In my opinion, while there is still farmland it [conversion] should be stopped, because when it is made into housing, we poor people cannot buy the housing. We don't have any way of earning income, especially when it's all housing that is constructed and not factories; but not all members of the family can work in the factories. When you're over-age you can't work, when you're under age you also can't work. If you haven't finished high school, if you don't have a strong backer, you also won't be accepted. If these remain as farmlands, then even when you're older you can still harvest and plant crops.... Because it's like this. The usual thinking of the people here is that, even when they disapprove of it, but the mayor likes it, their wishes will not be respected.
>
>> (Farmer in Mulawin, 1995; translated from Tagalog)

Several points emerge from these quotations. Firstly, and most obviously, agricultural land provides income and livelihood for farming households. Secondly, however, farmers also feel a strong attachment to the land on which they work, and their resistance to land conversion is as much visceral as it is a valuation of their source of livelihood. Finally, this individual resistance is not simply rooted in a financial and emotional loss, but also in the subsistence economy of the farming household that rice cultivation supports. Resistance, then, is based on three factors: the loss of a source of income in the formal economy; emotional attachment to the land; and, the removal of a subsistence food supply.

But the quotes above also indicate the reasons why such resistance has not been articulated in a collective fashion in Tanza. Except for a few cases to be discussed later, most instances of agricultural land conversion have passed through unopposed. Even where farmers had legal rights to block the conversion they have not done so. In part, this is clearly related to the marginal profits to be made from rice cultivation, described in chapter 4, and the substantial compensation payments that farmers receive for their eviction. But the quotes indicate that the embeddedness of these financial transactions in local social and political power relations also plays a major part in securing conversions. As chapter 5 suggested, the social context of land conversion places individual farmers in a position of culturally constrained behaviour towards their landlords and in politically impotent relations with local government officials. Farmers are 'ashamed' to contradict the (illegal) will of their landlords, with whom a personal, although unequal, relationship has been established over many years. Likewise, the local political machinery is all too often beholden to those with the most influence and deepest pockets, rather than functioning as an impartial regulatory framework. For these reasons, resistance has been limited largely to private complaints and regrets.

In a few instances, however, broader oppositional movements have arisen that contest the process of urban-industrial development being undertaken in the name of globalization. Although they frequently include the types of civil society organizations described earlier, such resistance is often closer in nature to social movements, which, according to Scott's definition, form

> a collective actor constituted by individuals who understand themselves to have common interests and, at least for some significant part of their social existence, a common identity. Such movements are distinguished from other collective actors, such as political parties and pressure groups, in that they have mass mobilization, or the threat of mobilization, as their prime source of social sanction, and hence of power. They are further distinguished from other collectivities, such as voluntary associations or clubs, in being chiefly concerned to defend or change society.
> (Scott, 1990, cited by Rodan, 1996: 18)

So-called New Social Movements (NSM) have been elevated to a hallowed status in much of the recent literature on the politics of postmodernism. They are characterized by an internationalized orientation, their subscription to values that challenge the existing social order, their decentralized, informal and often transient organizational structures, and their issue-based motivation (Rodan, 1996).

Movements with such characteristics have emerged in Cavite. Labour regulation, for example, has been a source of local protest. Cavite's non-union and anti-strike policies under Governor Remulla have precipitated unrest (Coronel, 1995).[12] The most vigorously contested issue, however, has been the conversion of agricultural land into urban and industrial uses. On occasion such resistance has erupted into violent confrontations. Examples have included resistance to the original construction of the Cavite Export Processing Zone in Rosario (1981), against an industrial estate developed by the Japanese Marubeni Corporation in Langkaan, Dasmarinas (1990), against a new dumpsite servicing Manila to be located in Carmona (1992), and against numerous other residential and industrial estates.[13] Some of these protests have achieved national and international exposure. The 'Langkaan Controversy', for example, led to the ousting of the Secretary of Agrarian Reform in the Aquino administration after he refused to sign a land conversion permit for the development (McAndrew, 1994). More generally, the conversion of farmland has been a point of heated debate within the Philippines, particularly among progressive groups concerned with the eviction of tenant farmers and issues of national food security.[14] Here, two examples will be highlighted where resistance in Cavite has been successfully mobilized.

The politics of power in barrio Amaya

The first example comes from a village near those described in chapters 4 and 5 and provides an illustration of some of the mechanisms of social movement resistance.[15] The village in question is Amaya, the second largest population centre (with 22,000 inhabitants in 1990) in the municipality of Tanza (see map 3). It was on 35 hectares of land in Amaya that municipal councillors invited an international consortium of investors to locate a 330 megawatt power station in 1992. In a resolution dated 28 September 1992, municipal councillors urged that the plant be located in Tanza indicating that it would uplift economic conditions and encourage other foreign investors to consider the town as a place to do business.[16] A few days later the municipal council passed an ordinance declaring the rice land in question to be officially rezoned as industrial. The project was symbolically launched at Malacañang Palace (the presidential palace) in Manila on 12 December 1992, with the enthusiastic support of President Ramos, who had invested much personal credibility in resolving the problem of frequent power cuts (a problem that was strongly dissuading potential investors from locating in the Philippines).

The Amaya plant was to be constructed at a cost of US$275 million by a consortium of investors, both domestic and foreign, who formed the Cavite Energy Corporation (CEC). CEC was, in turn, a subsidiary of Tradeinvest Asia Inc. – itself a subsidiary of Hong Kong-based Ace Indonesia Incorporated. The local broker was a Manila-based entrepreneur, but technological inputs and some capital were apparently coming from an Australian company. The plant would sell power to the Manila Electric Company (Meralco) for twenty-five years before transferring the facility to the government.

The first that local people heard about the project was a two-page newspaper advertisement in December 1992 declaring that ground-breaking ceremonies would be performed in a few week's time. Yet clearance from the Department of Environment's Environmental Management Bureau (EMB) had not at that point been obtained and was conditional upon technical soundness, including environmental considerations, and social acceptability. A highly simplistic environmental report had been prepared, without any reference to local conditions, and social acceptability had not been evaluated at all, yet over 100 families would have to be relocated for the project.

By the end of January 1993, local opposition was mounting as news spread of the consequences of the project. When EMB evaluators came to the municipal hall to assess social acceptability in February 1993, 7,000 Amaya residents marched from the village to demonstrate their opposition. The response from local political officials was to give assurances that the matter would be carefully evaluated, but to emphasize that the power plant was important in assisting industrialization through foreign investment. Municipal officials had declared the site as zoned for industrial use, but no maps or documents were available at the municipal hall to indicate this on an official land use plan. At the provincial level, the protesters received an even more dismissive response and experienced the wrath of Governor Remulla. A local leader of the protest movement described their encounter with the Governor:

> Some of the people that were with us recorded our conversation, our dialogue with the provincial governor. We were with some nuns and we were talking about the problem. And you know he got agitated, very very agitated. I never saw him like that. He said, in Tagalog, '*umuwi na kayo*', just like a father telling his infant child: 'go home!'. 'I'm your governor I know what I'm going to do, you go home'. He was with his bodyguards. Up in his office in Trece Martires. Several nuns were with us. There were about 300 people. This was the time we were already in the courts, a court was hearing our petition to have that [zoning] resolution nullified.

By September 1993, EMB approval had been secured by the consortium, leading to speculation that bribes had been paid at the highest levels in the Department of the Environment. It seemed that popular protest, still inscribed on many walls in Amaya with the words ' No to Power Plant', had achieved little.

It was, however, in mid-1993 that non-governmental organizations (NGOs) became involved, including the Rural Missionaries of the Philippines, Solidarity for Peoples' Power, The Southern Tagalog Alliance for Genuine Development Alternatives (ST-AGENDA), and the Philippine Environmental Action Network. As part of the Philippines' progressive movement, these organizations viewed the proposed development in Tanza not just as a socially unacceptable imposition on local people but also as a manifestation of the government's reliance on foreign capital and prioritization of industrial projects over agricultural development. They thus made connections between the specific project under dispute and the broader worldview that it represented.

The NGOs organized press conferences and media coverage and a corruption charge was laid against the Secretary of the Department of Environment and Natural Resources for issuing an environmental clearance without proper public consultation.[17] NGOs experienced in public protest coached local people in appropriate strategies. The groups' leaders stayed in Amaya during the most intense periods of protest and some, from more militant groups, indicated a willingness to take drastic action, including bombing, if construction of the plant went ahead. Other strategies of protest also developed during the latter half of 1993. On one occasion, opposition leaders learnt that the President would be passing over Tanza in a helicopter to attend the funeral of a murdered mayor elsewhere in Cavite. Local people were mobilized and assembled on the proposed power plant site to form, six-abreast, a giant human chain spelling the word 'NO'.

The NGOs also started to mobilize their international networks. Through contacts in Australia, the project's partner there was investigated and a protest was lodged with the Prime Minister's Office. A letter arrived in Tanza indicating that the Australian government would investigate the affair. Oppositional groups were also mobilized in the Netherlands and elsewhere, and the NGOs ensured that foreign visitors, especially those from influential countries such as Japan, saw the site in Amaya.

All these pressures led to delays in the project's commencement and as a result its prospective customer, Meralco, raised the possibility of cancelling its power purchase agreement with CEC. With the withdrawal of some investors, the funding package started to collapse. The CEC looked elsewhere for money and started negotiations with Energy Initiatives Incorporated, a subsidiary of the New York-based General Public Utility (GPU), a power supplier in New Jersey and Pennsylvania. But they too pulled out, and the project was left without sufficient funding. Since then, other sites in Cavite have been earmarked for power plant development and the Amaya site appears to have escaped such a fate.

The example provided by the Amaya power plant finds parallels elsewhere in Cavite, where local 'everyday resistance' and impromptu social movements have been translated into effective resistance when activated at larger scales and with broader agendas by NGOs. The other important feature of the Amaya protest was that it was supported by a broad cross-section of the local population. Even those without farmland being affected resisted because of the plant's potential environmental consequences in their barrio. The opposition was thus supported by landless labourers, tenant farmers, landowners and professionals alike. This unity of purpose across social groups and across scales would appear to be a precondition for 'local' resistance to globalization. Even so, it was still 'local' power structures, in the form of municipal, provincial and national government agencies, which the oppositional movement had to tackle, rather than the global developers *per se*.

Plate 11 The campaign that toppled the Kingpin: the Velsaco–Revilla ticket, with tacit support from 'FVR' ended Remulla's rule over Cavite

The Kingpin's usurpation – electoral resistance in Cavite, 1995

A second, and quite different, moment of resistance can be identified in the electoral defeat of Governor Juanito Remulla in the provincial elections of May 1995. As discussed in chapter 3, Remulla possessed a formidable political machine in Cavite and his electoral defeat was one of the major political upsets of the 1995 local elections. The reasons for Remulla's demise are instructive and can be identified in both his relations with the national scale of politics and with a disaffected electorate within his province. Only when these scales were united against him was the 'kingpin' usurped.

Although a member of the same coalition as the President, Remulla had earned the enmity of Fidel Ramos during the 1992 presidential election, when he chose to support the rival candidacies of Senator Ramon Mitra and Eduardo 'Danding' Cojuangco. Remulla had orchestrated an improbable vote of zero from the party's provincial branch for the Ramos candidacy, although in the presidential election itself, Ramos won in the province.[18] This, along with Remulla's reputation as a 'strongman' with strident policies towards land conversion and labour laws that embarrassed the national administration, fuelled Ramos' determination to see him replaced.[19] The President found a Caviteño of sufficient stature to oppose Remulla in the form of Epimaco Velasco, a native of Tanza, and the Director of the National Bureau of Investigation. Velasco's candidacy was given glamour and popular appeal by the addition of Bong Revilla, a twenty-seven-year old action-movie star and son of a Caviteño senator (see plate 11). The Velasco–Revilla organization also fielded a full slate of mayoral candidates across the province, drawn from the upper classes of Cavite's towns.

Remulla's powerful alliances within Ramos' Lakas–Laban coalition meant that the President could not openly endorse Velasco's candidacy or campaign for him, but voters were left in no doubt as to the President's feelings. His initials – FVR – were used in the names both of Velasco's *ad hoc* political party, 'Forward with Vitality and Reform', and his declared funding source, 'Friends of Velasco and Revilla'. Powerful backing was also evident in Velasco's well-organized, well-funded, and well-armed campaign. Most important, however, were the issues on which Velasco chose to base his campaign – putting a stop to the conversion of irrigated agricultural land, and asserting the rights of factory workers. Velasco argued that the provincial government had gone too far in allowing the wholesale conversion of valuable agricultural land and had not focused land conversion on upland areas of less importance. He also denounced the repression of labour rights in the denial of minimum wages and union organization, particularly in the Cavite Export Processing Zone.[20] These messages were carried in a vigorous and well-funded barrio-level campaign.

Election day, 8 May 1995, saw Velasco finishing with 78,163 votes and Remulla with 55,638; Revilla also won by a similar margin.[21] Two features stand out as quite startling in this result. The first is simply that Remulla's political machinery could let him down in such a way. The second is that his defeat included rejection by those towns that had 'benefited' the most from his policies of globalized development.

The governor had a reputation for being able to deliver votes (through his many patron–client relationships across the province, or simply through direct bribes and intimidation), which had earned him many powerful allies in the Senate and Congress. And yet, when his political career depended on it, he was unable to deliver votes for himself. The reason for this appeared to be the formidable array of resources available to the President's candidate. As a former director of the NBI, Velasco had access to sufficient 'gun and goon' power to neutralize any threat from Remulla of strong arm tactics.[22] In effect, therefore, the election was contested on the basis of the issues which Velasco laid out in his campaign – a rarity in Filipino local politics.[23]

These issues were highlighted by the geography of the election result. Remulla, a native of Imus in the north of the province, won only by the narrowest of margins in his own home town and lost in most of the other lowland towns where his 'revolution' of industrialization and urbanization had been most effective. In these areas, it seems, the electorate was dissatisfied with the opportunities of employment stemming from global capital investment. Rather than respond positively to Remulla's development strategy, the electorate reacted against the anti-union policies that he had consistently and openly held. In short, the 1995 election result suggests that widespread opposition did exist to Remulla's policies, but it was suppressed until a candidate emerged who could neutralize the electoral tactics usually employed.[24] Thus a rare occurrence was witnessed in Filipino politics (and a perhaps unique occurrence in Cavite politics) – a sitting strongman was removed by a popular electoral vote.

While acknowledging this fact, Remulla remained unrepentant. A few days after the election he commented in the press: 'I really find it ironical that those who benefited from my administration's industrial peace policy were the ones who caused my downfall'.[25] The governor maintained that his political practices were legitimate given the need to attract foreign capital and industrial investment to the province.

Early indications suggested that Velasco's election brought a very different tone to the provincial administration: a billboard erected by Remulla's government on the highway into Cavite from Manila announced the province's 'Second Revolution' of industrial development and its 'Peace and Productivity Zone'; a few weeks after Velasco was elected, the sign was painted over and replaced with the slogan 'Touch the Lives of the Less Fortunate: Pay your Taxes'. At the same time, however, the possibility of significant change in Cavite's political economy would seem to be unlikely given the election of nineteen out of twenty-three mayors, three congressmen, and many provincial board members who stood as Remulla's candidates.[26]

Two conclusions can be drawn from this episode of successful electoral resistance to land conversion and labour repression. Firstly, it was a coalition of powerful candidates together with disaffected farmers and factory workers that succeeded in toppling the figurehead of Cavite's globalized development. As in the case of the Amaya power plant, it took an alliance such as this, cutting across social and economic groups, to form an effective opposition. Secondly, it was the combination of different scales within this opposition – from the national scale of presidential support to the *barangay* scale of Velasco and Revilla's political campaign – that permitted 'local' opposition to overcome those political leaders who had derived their power from mediating the relationship between 'local' and 'global' scales. In other words, only when Remulla was squeezed from both above and below was his power to mediate the between scales successfully challenged.

Ambivalence and ambiguity in resistance to globalization

Thus far we have seen how intellectuals and activists have articulated oppositional discourses to counter the dominance afforded to globalized development, and how social movements and NGOs have synthesized popular discontent into a few but notable victories. What these examples do not imply, however, is evidence that 'local people' resisted the exigencies of globalization, as some would suggest (for example, see Seabrook, 1996). The situation is more complex than that.

Firstly, there is considerable ambivalence towards the changes wrought by globalization in Cavite, and the Philippines more generally. Here we might usefully invoke Massey's (1994) suggestion that there exists a 'power geometry' to globalization, meaning that its impacts are differential experienced according to various social axes. The existence of such a differential experience of globalization implies that there can be no unified 'local' response to the process because it means different things, both materially and culturally, to different people. The economic 'power geometry' produces a series of axes on which those variously located in local power structures are positioned. Local politicians, large landowners and property developers (and the three groups overlap significantly) are the prime beneficiaries of urban and industrial development, followed by tenant farmers and their families who receive compensation payments. Various groups also benefit directly, such as those with employment based on manufacturing industries or the construction sectors. Others will derive less direct benefits if, for example, they are involved in the transportation, service or retail sectors.

Entwined with this economic power geometry are cultural axes of access, exclusion and change in the local process of globalization. At a basic level, various aspects of individual identity shape access to globalization. As the quote earlier in this chapter made clear, employment at local factories is selective on the basis of age, gender, education and political connections. But in a more complex way, cultural identities are reworked. Gender identities and youth subculture are transformed in often liberating ways, as described in chapters 4 and 5. At the same time, changing attitudes to work among young people mean that land conversion is less an issue to them than it is to their parents' generation. In this way, globalized development can open new spaces for the construction of identity, especially among younger people, and these spaces have both emancipatory and exploitative dimensions.

In Cavite, then, globalization is experienced in different ways by different people, and even in different ways by the same people, as both costs and benefits are weighed. In only a few instances can there be seen an absolute loss resulting from land conversion and industrial development. Most clearly this is the case among older, uneducated migrants who have come to Tanza to work in agriculture. Without access to the benefits of industrial employment, without land of their own for which to be compensated, without political connections through which to secure new opportunities, and often without kinship networks to provide assistance, these people represent the poorest and most socially marginal group. For others, however, variously positioned within the power geometry and identity distinctions of globalization, ambivalence and acceptance are understandable reactions. It is for these reasons that we cannot easily talk of 'local' resistance to globalization.

It also important to recognize ambiguity in the process of globalization as it is localized in social and physical landscapes. For while the 'local' might again be pitted against the 'global', the actual mechanisms of power through which globalized development is experienced are decidedly local. Specifically, these power relations include the mechanisms for labour repression and land conversion described in earlier chapters. They operate at the level both of local politics and of everyday politics – through the activities of mayors and governors, and through the socially constructed relationship between landlords and tenants.

The broader point to draw is that to speak of 'local' resistance to 'global' forces is to misrepresent the context of change. Instead, the experience of farmers in Tanza shows that 'local' resistance is pitted against 'local' processes of change, because it is only through local structures that 'global' processes are manifested.

Thus, the dichotomous construction of 'local vs. global' does not faithfully represent the social, political and cultural context for experiencing globalization that has been described in several chapters of this book. Firstly, while there are many instances in which individual opposition is expressed, particularly by farmers in Tanza, their place within local power structures is such that they cannot activate this resistance in any meaningful way. Secondly, while the manifestations of globalization are explicitly harmful for some, for others the opportunities that arise, economically and culturally, are liberating. Moreover, households and even individuals may experience both positive and negative reactions. Thus the response to globalization is unsurprisingly ambivalent. Thirdly, where resistance to globalized development is successful, it is because local social and political relationships have been transcended, or 'dislocated', in a coalition of opposition.

Conclusions

This chapter has discussed some of the possibilities for resistance to the process of globalization. What emerges is the conclusion that speaking of 'local' resistance to the 'global' is an overly simplistic representation. The relationship of individuals to global flows is complex and ambivalent and does not necessarily create the grounds for grassroots opposition. Furthermore, where there is opposition, and the long quotes provided earlier indicated that it does exist quite passionately, it comes up against social barriers to protest and local power structures that have a vested interest in globalized development.

There have, however, been moments of successful resistance, exemplified by the Amaya case and the 1995 elections. These examples seem to point towards certain preconditions for successful oppositional movements. Firstly, diverse and fractured local experiences of globalized development must be united. Only when local structures of social and political power are transcended can the opposition that is contained by them become activated. Secondly, this transcendence of social groups must be matched by a transcendence of multiple scales of opposition. Whether this is brought about by the involvement of NGOs or a disgruntled President, it is a crucial part of resisting the power of those who promote and benefit from globalized development.

7

CONCLUSION

Several arguments have been woven through this book, surfacing through detailed empirical information. Here I will attempt to clarify what I believe the experiences of the people and places described have to contribute to a broader understanding of what globalization means. For while the case studies explored in this book have been fairly parochial, their implications extend beyond the processes of change in a few villages, or even in the Philippines more broadly. They also speak to the nature of globalization as it has been represented and imagined in popular literature, political rhetoric and academic work, and how this representation needs to be reformulated.

The politics of globalization

The first set of arguments concerns the politicized nature of globalization. Globalization refers to a set of material processes that are integrating social relations both in more intensive and more spatially extensive ways. Clearly, the studies presented in this book have been especially, although not exclusively, concerned with economic processes of integration. The existence and material importance of such processes is undeniable – experiences in Cavite showed the importance of flows of capital, people, commodities and technology for everyday life. But at the same time, the processes of globalization have been elevated to something that is more than the sum of their parts. Through political rhetoric, planning documents and academic analyses, globalization has become a discursive construction disembedded from its material realities. In particular, it has been deployed as an idea to project an imaginary landscape of the global economy, and the place of the Philippines within that landscape. This landscape is not a conventional one of topographic features and absolute spaces, but is instead a space of flows and networked relationships. The role of places in this imagined landscape is as nodes where disembodied flows of capital, technology and so on may (or may not) intersect. Furthermore, only through tapping into these flows is prosperity and development possible – it is, we are told, irresistible, inexorable and inevitable.

This representation implies the necessity of positioning places to be receptive to such flows. In the late 1990s this has been a 'necessity' that has come under threat as a regional and increasingly global financial crisis exposed the weaknesses and vulnerabilities of economies overexposed to fickle global financial flows. Whilst some political economies, notably Malaysia, responded by partially rejecting a globalist perspective through the introduction of capital controls, in the Philippines the crisis appears to have deepened political resolve to assert the supremacy of global space in the country's development strategy. On his first overseas trip as President in October 1998, Joseph Estrada visited Singapore to thank Filipino overseas workers for their contributions to national development and to address business and government leaders. His message was clear: 'We are pro-market and shall continue to be a market-driven economy. We are pro-investment and pro-competition and shall not impede the way to international integration'.[1]

Globalization, then, continues to be a powerful political talisman in the Philippines and is used to justify a set a specific economic development policies. In particular, the drive to attract export-oriented foreign direct investment has been at the forefront of national economic strategy. In practical terms this priority has several corollaries. Firstly, it implies (and has resulted in) a reduction in the importance of agriculture in national development strategies. Rather than giving priority to developing infrastructure such as irrigation systems or farm-to market roads, it has instead been given to the needs of foreign investors in the manufacturing and service sector. This has been particularly evident in the productive agricultural areas in the national core region, where nationally important rice growing areas have been converted to urban and industrial uses.

The changes experienced in the national core, and especially in Cavite, also highlight political issues of spatial equity. While chapter 3 showed that a variety of factors can be used to explain the skewed nature of the Philippine space economy, priority has clearly been given to core areas in terms of infrastructure provision and investment promotion, assuming that a process of spatial trickle-down will eventually occur. Political choices, then, are made in the name of globalization with respect to both sectoral and spatial development strategies.

The third political dimension of globalization lies in the provision for the needs, or perceived needs, of global (foreign) capital. This has been highlighted in this study with respect to the management of two key factors of production: labour and land. The creation and regulation of an industrial labour force in a province that was almost exclusively agricultural just a few decades ago have been central to Cavite's appeal to foreign capital. The constant emphasis on 'industrial peace and productivity' and the various strategies used to quash attempts at unionization have been vital in promoting the province as an appealing 'node' in a global network of investment opportunities. Similarly, the management of land conversion for industrial and ancillary urban uses has meant that a stream of industrial sites has been assured at prices acceptable to investors. Thus, while proximity to Manila and connections with telecommunications, air transport and port facilities have assisted Cavite's drive for industrialization, it has been the ability of its political leadership to assure that the practical needs of investors are met that has ensured its status as the country's pre-eminent industrial hub.

The involvement of local politicians – mayors and governors in particular – in the process of land and labour management highlights a final political dimension to globalized development. The ability of such figures to act as go-betweens in the entry of foreign capital to their jurisdictions has enabled them to entrench their political power through the financial benefits accompanying such development. But this entrenchment occurs not simply through the lining of pockets to finance political machines. It has also occurred through the discursive self-representation of power holders as the intermediaries in bringing the benefits of rapid industrialization to their people. Thus, political bosses can construct an image of the global, mediate local experience of global flows and, at the same time, legitimize local practices and their own authority with reference to same discourse of globalization. Power, Rafael noted in an abstract sense is 'the capacity to lay claim over the site of circulation and thereby broker the exchange between the inside and the outside' (Rafael, 1995: xix). In Cavite, we can observe this process in action.

Embedding and embodying globalization

The implications of globalization in the constitution of local power relations brings us to the second set of arguments that have informed this study. These relate to the ways in which the material processes of globalization are actually embedded in social processes at other scales. In other words, how do social processes (economic, political, cultural and environmental), embedded in places at multiple scales, mediate and construct a particular experience of globalization?

Chapter 2 started with an account of the ways in which 'local–global' relations in the Philippines have been constructed over time at a national scale. This historical account traced the emergence of a complex political economy in the Philippines that incorporated distinctions of ethnicity, class, family ties and social status. Over the course of colonial and post-colonial history, the vested interests of diverse domestic constituencies made the relationship between the Philippines and the growing world economy an arena of contestation. By the late 1960s, however, a strong constituency oriented towards foreign-investment driven, and export-oriented, development had emerged. Political changes within the country in the 1970s, most significantly the imposition of martial law, resulted in a drive to attract foreign investment in export manufacturing, but contradictory policy tendencies (reflecting the interests of competing constituencies), coupled with a personalistic regulatory environment, meant that little was achieved in terms of attracting foreign investment flows to locate in the country. The removal of Marcos, however, lead to more concerted attempts at globalized development and by the time of the Aquino and Ramos administrations the political consensus around this strategy had solidified. The Ramos government in particular deployed a rhetoric and liberalized policy framework firmly grounded in the discourse of globalization. As a result, indicators of foreign investment, exports, and economic openness took a sharp upward turn in the mid-1990s. The Philippines' 'place in the world' had become firmly constructed as a node in a global space of flows and an explicit rhetoric of globalization had emerged to legitimize this position – seen for example in place-marketing campaigns based on selling the 'strategic' location of the country in a global space of flows. Nevertheless, such a construction must be seen as a contingent reflection of the interests of powerful groups and actors within and outside the country.

It is, however, at a sub-national scale that the embeddedness of globalization is most apparent. At provincial, municipal and village scales, globalization in Cavite has been shaped by a system of contemporary political-economic power relations deeply rooted in its colonial history. This mediation of global flows by local social relations both reflects and reinforces this power. By representing Cavite as a node or 'site' in a global network of investment flows, local politicians have been able to justify a set of practices with regard to two key factors of production – land and labour – in order to satisfy a third, capital. But the process is not one of unrelenting structural power on the part of 'capital'. Instead, industrial investment has been incorporated into the local political economy on terms dictated by, and beneficial towards, those in command. The processes of labour market transformation and agricultural land conversion in particular were shown to be driven by globalized development but moulded by powerful local figures.

These processes of labour market transformation and land use conversion were explored in the villages of Bunga and Mulawin respectively. In each case these processes cannot be understood solely as globalization inscribed on the local social or physical landscape. Both represented the integration of globalized development, manifested in the nearby export processing zone, with local social relations. The scarcity of labour experienced in agriculture, for example, is not *just* a result of massive increases in employment at the factories in the zone. It also reflects: the labour needs and divisions of labour inherent in the human ecology of rice cultivation; the social relations of tenancy; the cultural significance of rice; and the changing attitudes towards farm work. All of these locally constituted factors determined the ways in which globalization was experienced, and responded to, in a village such as Bunga.

Locally constituted social relations are also critical in understanding the loss of agricultural land in Mulawin. The nature of tenancy arrangements, the legal and political context of land use regulations, and the environmental conflicts between rice ecology and urban development, all contributed to the conversion of agricultural land even where it is destined to simply lie idle. To leave land vacant in this way cannot be rationalized on purely economic grounds; it is only in the light of these social, political and environmental relationships that it can be explained.

Globalization is also constituted even at the scale of the individual. Various axes of social and cultural distinction have been shown to play a part in determining how individuals relate to globalized development. Age and gender, for example, form fundamental filters to employment opportunities in local factories, as recruiters seek out young, single women. Masculinity is constructed as being too prone to insubordination and laxity in the workplace, while older men and women would also resist factory discipline and require a more 'liveable' wage. Formal educational attainment too effectively excludes many of the older generation, as employers demand high school graduates. These 'identity barriers' are breached in many cases, where falsified birth certificates and high school diplomas and concealed marriage and motherhood have allowed women to gain employment in otherwise unlikely circumstances. In general, however, the filters in the recruitment process serve to limit employment to those favoured by factory managers.

Other axes of identity also play a major role in determining an individual's experience of globalized development. Most importantly, it is on the basis of political-economic differentiation, particularly with respect to land ownership and access, that the benefits of land conversion are distributed. Landowners take the lion's share of the value added in conversion, while tenant farmers can also negotiate a substantial compensation package. The conversion process is, however, closely tied in with political connections, both locally and nationally, and so affiliations based on factional allegiance and kinship ties determine access to such channels. The most disenfranchised group in the municipality, with respect to both class status and political connections, are new migrants, many of whom work as agricultural labourers. They live without legal access to land (and therefore the benefits of its conversion), without kinship and political networks through which opportunities arise, and in some cases face discrimination and social exclusion.

These, then, are some of the major axes of individual identity along which access to globalized development is distinguished. But cultural identities do not operate in a simple, static way; identity and subjectivity are not solely created from locally constituted and immutable categories onto which the dynamism of globalization is superimposed. These categories enter into a reflexive relationship with processes whose provenance is beyond the scale of the village, town or country. In other words, at the same time as various axes of cultural identity are active as 'filters' in the subjective experience of globalization, the metaphor of the 'filter' is unsatisfactory because these categories themselves are reworked in the very process they help to shape. Thus attitudes to work and gender roles, in particular, have undergone significant transformations in recent years in the areas studied in Cavite. Cultural identities are reworked, but in ways that incorporate elements of both the 'local' and the 'global' – a process better captured, perhaps, through such notions as 'hybridization' and 'in-betweenness'.

Massey's 'power geometry' of globalization, then, is useful in understanding the differential access to global flows. Such flows, in their very nature, are not disembodied phenomena but refracted by local social relations in such a way that experiences of globalization both reflect and reinforce power (Massey, 1994: 62). Thus, the ways in which individuals experience globalization is defined by their positioning in the local political economy. However, local geometries of power and identity are also reworked in the process of globalization, meaning that globalized development neither reworks 'locality', nor is it simply mediated by local social relations; instead, it is constituted or embedded in multiple scales.

Relativizing the global scale

Implicit in the discussion of the 'embeddedness' of globalization is the argument that the 'local' is no more meaningful as a spatial category that the 'global'. While the impersonal and disembodied global logics that were reviewed in chapter 1 have been shown to be rather more localized and 'embodied' than many writers allow, this localization occurs at *multiple* scales. In other words, the 'places' in which globalization is embedded are simultaneously national, provincial, municipal, village, household and individual. Social processes which activate globalization operate at, and between, each of these scales. Thus, for example, we have seen how the political power brokers of Cavite could not exert such influence over the globalizing labour process and the land market without ties to both national level authorities and municipal and village level agents. Similarly, defining development policy in the national arena has been a process of brokering vested interests at that scale with the broader context of colonial powers and latterly international creditors. The process of globalization, then, is activated and experienced in social relations operating at multiple and connected scales.

The corollary of this argument brings us back to the theoretical discussion of relativized scale in chapter 1. Scales cannot be hierarchically arranged, with social processes at larger scales determinative over those at smaller scales. Nor can scales be reified in suggestions that the 'global' is in some sense an actor or influence in its own right. And yet this is exactly the construction of scale that is embodied in the discourse of globalization. It incorporates a notion of the global scale as a privileged domain of analysis. By representing places as nodes in a global space of flows, alternative political options – most notably some form of economic nationalism – are deferred in the face of the determining inevitability of processes at a global scale, for which no one can be held accountable.

Instead, the account provided here suggests that analysis of the material changes in the social and physical landscapes of Cavite requires a relativized view of scales in which the global does not imprint itself upon a passive terrain. Rather, processes of rapid capitalist development as witnessed in Cavite are constituted in multiple scales of power relations. This, I believe, explains why the frustrations of the farmers quoted in chapter 6 have only occasionally found an effective voice. It is not because they are powerless against the impersonal forces of globalization, but rather because they are disempowered within the context of local social power relations. It is those wielding power who can construct globalization to legitimize their actions and further their interests. Thus it is only when these local power relations are transcended or dislocated, for example through the involvement of NGOs in the Amaya dispute, or the intercession of Presidential influence in the 1995 gubernatorial election, that disempowered voices are heard.

Rethinking globalization

Both political and academic conclusions emerge from these arguments (although hopefully the two are not mutually exclusive categories). Firstly, globalization, in this case as flows of industrial investment, must be seen as an inherently localized process rather than as a universal and homogenizing force. It is through local political and social structures, discussed in this case at the provincial and municipal levels, that globalized development is constituted. From there, other scales spiral outwards. The implication for resistance to the changes that globalized development brings is that it must start with local social processes and power relations rather than necessarily focusing attention on the global scale. Thus, while the activities of NGOs discussed in chapter 5 are laudable, to look for alternatives to globalization at the global scale misses the importance of other scales in the social processes that are being identified as 'global'. The examples of 'everyday' resistance to globalization in the same chapter highlighted the fact that only when they operated across scales were such movements successful in attaining their goals.

The second conclusion that emerges is that rethinking globalization requires the recognition of its discursive construction. If the argument is accepted that processes of globalization are activated at multiple scales, then the use of globalization to legitimize political practices represents a discursive *invocation* rather than a material *determination*. Political choices are deferred to a constructed global scale that is represented as apolitical and unavoidable. Here we have seen instead that the social relations operating at multiple levels of the 'local' are where the experience of globalization is created.

How, then, might this political discourse be countered? At an intellectual level, the account provided in chapter 2 provides an 'archaeology' of globalization in the Philippines and in so doing demonstrates its historical contingency. Rather than an inevitable contemporary context for development, globalization begins to look more like a contested and constructed discourse. The implication is that alternative constructs, alternative imaginations, are required in order to subvert the dominance discourse of globalization and the power relations associated with it in the Philippine context. Chapter 6 showed that there is indeed a rich source of such alternatives circulating in the Phillipines that deserve a wider audience. These alternative discourses do, however, need to address the embodiment and embeddedness of globalization in contexts at multiple scales, starting but not ending with the most immediate context of power relations.

What this produces is a perspective on the material processes of globalization that views them as embedded, mediated, negotiated and resisted in places at multiple scales. This refutes arguments that imply an 'end of geography' brought on by a global space of flows. The processes of globalization are not inexorable logics, but are received, interpreted, accommodated and adapted in particular ways in different places. Where, then, does this leave the study of globalization and social change in places like Bunga or Mulawin, Cavite, or the Philippines? I would argue that it should be situated in, and between, each of these multiple places, because it is precisely there that the material processes of globalization are embedded in local power relations: power relations that discursive constructions of globalization all too often serve to legitimize.

APPENDIX: METHODOLOGICAL ISSUES

The main periods of fieldwork upon which this study is based were in mid-1994 and the between January and August 1995. Two brief return visits were made in 1998. The research employed a variety of techniques in 1995 and this appendix will describe each in turn.

Secondary sources

The data and literature gathered at various offices and institutions in the Philippines were broadly related to agriculture, urban development, industrialization and economic development policy. In some instances, just secondary material was gathered at government agencies and other institutions, but frequently I was also able to interview key personnel. While my status as a white, male, English/Canadian researcher created obstacles to understanding in some senses, it was also an enormous benefit in dealing with officials at governmental and non-governmental offices. An 'outside' identity frequently opened doors that would be closed to those within the local social hierarchy. This meant access to senior politicians, officials and academics was possible, while at the same time providing enough curiosity value that farmers, storekeepers etc. were often keen to engage in discussions. My association with the prestigious University of the Philippines also assisted in facilitating access to the bureaucracy.

Interviews

During the course of my research I conducted taped interviews with over seventy individuals, lasting from half an hour to three or four hours. These were predominantly with members of farming families in the villages of Bunga and Mulawin, but also with other residents of Tanza. Although a few interviews were conducted in English, most were in Tagalog. Interviews were transcribed a few days later in abbreviated form – taking notes in English for factual information, but transcribing complete passages of a more qualitative nature. The quotes used in this book are drawn from these interviews and have been left as fairly literal translations.

The majority of interviewees were members of farming families in the two villages.

In addition, however, I also sought out the following groups: workers at the CEPZ, landless agricultural workers, developers, agricultural technicians, youth leaders, teachers, and older people both in villages and in the town proper with a good recollection of local history. In addition to the taped interviews I also spoke to numerous others in less formal circumstances and recorded our discussions in written form.

Almost all of the interviews were conducted in the respondent's house, but occasionally in a field or other place of work. While guided by a series of predetermined, but open-ended, questions, most interviews went off on tangents determined by the interviewee.

Surveys

Three surveys were conducted during the course of my research in Tanza. The first survey instrument was a short (two pages) household questionnaire used to gather socioeconomic information in Bunga and Mulawin. The survey was written in Tagalog and tested on a few households before full implementation. It requested details of family members, educational attainment, migration, sources of livelihood (primary and others), current or returned overseas contract workers, responsibilities for household chores, and ownership of household appliances. The survey was devised only after a period of open-ended interviews which allowed me to develop questions that would be relevant and acceptable. In both Bunga and Mulawin, the bulk of the surveys were conducted over two or three days, with the help of my research assistant and several volunteers from the *barangay* youth groups (*Sangguniang Kabataan* or *Kabataan Barangay*), who were given a short briefing and training session.

In Bunga, we were able to complete 230 surveys out of a total household population of 260 (a coverage of about 90 per cent). In Mulawin, 205 surveys were completed but this represented only about one third of the total household population. The latter was not a random sample but represented the core area of the village which includes several housing developments of various ages. Two large developments which are technically within the village boundary were excluded because they are separated from the main part of the village and because it was necessary to keep the survey down to a manageable size.

The second survey was aimed specifically at farming households and sought information on the economic and technical details of their farming activities. Like the household survey, this questionnaire was also designed after a period of open-ended interviews, but was administered exclusively by my research assistant and myself. In Bunga we covered forty-three of the sixty-five tenant farmers, and in Mulawin nine of the thirteen who remain.

The third survey was administered within the Tanza National Comprehensive High School. The purpose of the questionnaire, which was completed during classes, was to gauge the family backgrounds, attitudes and aspirations of young people. These data might have been gathered in more detail through in-depth interviews, and while this technique was also used, many teenagers were uncomfortable with interview situations or open-ended survey questions. The brief questionnaire was therefore designed in consultation with the school's counselling office and consisted mostly of short answer responses. After a pre-test with 10 students, the survey was administered to 136 final year students (mostly aged 16). Little of this survey is used here: it is reported in more detail in Kelly (1999c).

Mapping

Hand-drawn sketch maps were elicited from several local people and were then incorporated into maps of the two villages which I drew myself based on observations. They feature all houses – included those omitted from the household survey – together with institutional buildings, physical characteristics and general land use patterns for June 1995.

Analysis

The household surveys and Bunga farming surveys were entered into the Microsoft Access relational database management system. This software allowed several data tables to be compiled relating to personal characteristics, household characteristics and farming practices. Cross-tabulations could then be generated both within and between these data tables to provide descriptive statistics for a wide range of socio-economic characteristics in Bunga and Mulawin. Access was also used for some secondary data gathered from government offices, for example on industrial establishments in Cavite. For data sets with smaller populations or more limited uses – such as the farming survey in Mulawin or the high school survey – manual calculations were performed to generate quantitative output.

NOTES

1 INTRODUCTION: PUTTING GLOBALIZATION IN ITS PLACE

1 Quotes drawn from President Fidel Ramos' speeches on a Philippine government internet site: http://www.philippines.gov.ph:80/mhp/textversion/ulatjul96.htm12; and 'Ramos thanks senators for summit support', *Philippine Daily Inquirer*, 3 December 1996, 2.
2 Appadurai's (1990) five 'scapes' – finanscapes, ethnoscapes, mediascapes, ideoscapes and technoscapes – which refer to accelerating international flows of capital, people, images, ideas and technologies, are widely cited as capturing the multiple dimensions of globalization.
3 Empirical examples are provided in books such as Corbridge, Thrift and Martin (1994), and Dicken (1998).
4 The notion of the production, and the politics, of scale has been explored by Neil Smith over the last few years (1992a, 1992b, 1993, 1996; Herod, 1997; Smith and Katz, 1993; Swyngedouw, 1997). This work draws on ideas of the production of space (Harvey, 1989a, 1990; Lefebvre, 1991), and on notions of metaphor in geographical theory (see Barnes, 1996).
5 The concept of embeddedness originates in work on the sociology of economic institutions (Granovetter, 1985). Granovetter defines embeddedness as the argument that 'behaviour and institutions [of economic action] . . . are so constrained by ongoing social relations that to construe them as independent is a grievous misunderstanding' (1985: 482). In recent years this point of view has been influential in economic geography through institutionalist and network approaches to economic entities (Amin and Thrift, 1994; Dicken and Thrift, 1992; Thrift and Olds, 1996; Yeung, 1994). More generally, a concern with the embeddedness of economic change in local social relations can be identified in Doreen Massey's work on spatial divisions of labour (1984, 1994), the 'localities' research that it later inspired, and in other non-essentialist theories of economic geography (Barnes, 1996).

2 THE PHILIPPINES AND THE GLOBAL ECONOMY: CONSTRUCTING A PLACE IN THE WORLD

1 Cited in *Philippine Daily Inquirer*, 8 May 1995: B10.
2 'Trade chief says new tariff program will boost economy', *Philippine Daily Inquirer*, 24 July 1995; 'Government to draw up long-term plan on investment promotion', *Philippine Daily Inquirer*, 31 July 1995; 'RP draws P2.8-B aid as economic reforms praised', *Today*, 24 July 1994; 'Singson says IMF pleased with RP performance', *Philippine Daily Inquirer*, 27 June 1995; 'WB official expects brighter economic prospects for RP', *Philippine Daily Inquirer*, 28 June 1995; 'IMF vows support for RP; money targets finalized', *Philippine Daily Inquirer*, June 1996.

NOTES

3 'FVR: Roll out carpet for foreign investors', *Philippine Daily Inquirer*, 25 July 1995.
4 Reproduced on a government internet site: http://www.philippines.gov.ph:80/mhp/textversion/ulatjul96.htm12
5 Examples can be found in the brochures entitled *The Philippines: Back in Business in the Gateway to Asia and the Pacific* (NEDA 1995b), and, *The Philippines: Your Competitive Advantage* (Board of Investments, n.d.).
6 *Far Eastern Economic Review*, 5 November 1998: 66.
7 For two accounts of the Philippine experience of the crisis see: Jurado (1998) and Intal and Medalla (1998).

3 GLOBALIZATION AND THE PHILIPPINE SPACE ECONOMY: PATTERNS, PROCESSES AND POLITICS

1 The Central Luzon and Southern Tagalog administrative regions also include some more peripheral areas – for example, the island of Palawan is part of Region IV, Southern Tagalog. For government statistics, however, these two regions together with the National Capital Region of Metropolitan Manila best reflect the core of Manila and its immediate surroundings. Finer geographical patterns within this core will be highlighted later in this chapter.
2 The relevant institutions here are: (1) the Philippine Economic Zone Authority (established in 1995, replacing the Export Processing Zones Authority) which directly administers several government-owned zones and oversees a growing number of private 'ecozones'. To locate in an ecozone, a foreign owned firm must export 70 per cent of its output and can then avail of certain financial incentives. (2) Non-export firms, or those wishing to locate outside of a formal zone, can register with the Board of Investments. If their activities fall within the broad 'priority' areas established by the annual Investment Priorities Plan they may also receive incentives. (3) The Subic Bay Metropolitan Authority and the Clark Development Corporation independently register investments and grant incentives within their jurisdiction.
3 Employment generation is used in this section as an indicator of new economic activity. This allows problems relating to project cost figures, which cover approved rather than actual investment, to be bypassed. It also highlights more effectively the actual local impact of investment.
4 Various authors have provided empirical studies of the development of desakota regions around cities in Southeast Asia. On Jakarta, see Douglass (1989), Firman (1995, 1996), Henderson *et al.* (1996), and Leaf (1994, 1996). On Bangkok, see Douglass (1995), Greenberg (1994), McGee and Greenberg (1992), Parnwell and Wongsuphsawat (1997), Ross and Poungsomlee (1995), and Setchell (1995).
5 Data cited in *Business World*, 23 June 1997.
6 As Fuchs and Pernia (1987) point out, however, firm-specific determinants of location may vary by sector. The example of Texas Instruments locating in the Baguio EPZ for climatic reasons is a case in point.
7 In addition to Remulla and Marcos, the University of the Philippines Upsilon fraternity also included among its members Benigno Aquino and the current congressman for the second district of Cavite, Renato Dragon. Dragon is now rumoured to be a potential candidate in the next gubernatorial election.
8 Public sector opportunities include the granting of contracts and other revenue dispersal; in the private sector, many elected officials have business interests in the construction industry and recruitment for factory jobs and overseas contract work; and, in the 'informal' sector, there are widespread allegations that politicians are involved in illegal gambling (or *jueteng*) syndicates.

NOTES

4 THE GLOBALIZING VILLAGE I: LANDSCAPES OF LABOUR

1 Reprinted in 'The very best for Cavite', by J. R. Remulla, *Philippine Daily Inquirer*, 22 May 1995.
2 'Who are [*sic*] behind this syndicate?', *Philippine Daily Globe*, 18 October 1990.
3 'Government sacrifices workers to attract investments', *Today*, 23 February 1995.
4 'Workplaces in Cavite perilous, report says', *Today*, 28 February 1995.
5 'Workers hit Korean owner of Cavite firm', *Philippine Daily Inquirer*, 4 May 1995; 'Cavite labor officials deny no-strike policy', *Philippine Daily Inquirer*, 9 May 1995
6 'Workplaces in Cavite perilous, report says', *Today*, 28 February 1995.
7 'Give back P20 to Cavite workers', *Philippine Daily Inquirer*, 7 July 1995.
8 'Remulla scored for union busting', *Philippine Daily Inquirer*, 1 May 1995.
9 'Export zone wage violators bared', *Philippine Daily Inquirer*, 21 June 1995.
10 'Antilabor setup in export zones scored', *Today*, 22 February 1995; 'Remulla scored for union busting', *Philippine Daily Inquirer*, 1 May 1995; 'Unions want Remulla probed', *Philippine Daily Inquirer*, 21 March 1995. See also Coronel (1995).
11 Historical dates and figures are drawn from oral histories collected during interviews with villagers in 1995.
12 After the completion of the principal period of fieldwork for this study, the road to Bunga was resurfaced in December 1996. Transportation services to and from the village are therefore greatly enhanced. It will be interesting to see how this greater connectivity – the creation of a new transactional space – will affect development in Bunga in the years to come.
13 These reasons relate to the structure of tenancy established during the Spanish colonial period, described in chapter 3.
14 An extensive literature exists on the roots and practices of tenancy in the Philippines. See for example Takahashi (1970), Wolters (1984) and Kerkvliet (1990).
15 A *cavan* of rice is generally taken to be the equivalent of approximately 50 kgs, although Wolters (1984) provides an estimate of 44–6 kgs, depending on whether or not the rice has been dried. It is usual for unhusked or unmilled rice to be measured in *cavans*, but milled rice (i.e. ready for cooking), or *bigas*, to be measured in kilos.
16 Unpublished 'Masterlist of Rice Farmers in Tanza, January 1995', Municipal Agricultural Office, Tanza, Cavite. In my own survey, forth-three of sixty-five farmers were surveyed in Bunga. In Mulawin nine out of thirteen remaining farmers were surveyed.
17 This process has been documented elsewhere in Cavite by McAndrew (1994).
18 Putzel (1992) and Riedinger (1995) provide extensive accounts of agrarian reform programmes in the Philippines.
19 CLTs are amortized through regular payments to the Land Bank and on completion result in the transfer of land ownership to the cultivator.
20 The impacts of green revolution technology in the Philippines are documented by Boyce (1993) and Feder (1983).
21 Vegetables and other crops grown in addition to rice include: Chinese cabbage (*petchay*); onion; okra; watermelon (*pakwan*); garlic (*pawa*); string beans (*sitaw*); cucumber; tomato (*kamatis*); balsam apple (*ampalaya*); mungo; patola; guava (*bayabas*); eggplant (*talong*); bell pepper; baguio beans; *sampaguita* flowers.
22 A locally made hand tractor would cost 35–40,000 pesos.
23 See note 21 for a list of commonly grown vegetable and fruit crops.
24 As table 3.6 indicated, by early 1998, this had risen to 54,141.
25 Literal translations are from the Tagalog–English dictionary, compiled by Leo James English (1986) and published by the National Bookstore in Manila. The contextual meanings were derived from discussions with my research assistant, Ms Berna Javier.

NOTES

26 The original use of the word was with reference to groups of criminals shipped to Manila for imprisonment. The closest equivalent in English would probably be 'gang' or 'fraternity'.

5 THE GLOBALIZING VILLAGE II: THE PHYSICAL LANDSCAPE

1 The 1995 figure is from my own survey and fieldwork; the 1989 figure is from unpublished data prepared by the Department of Agriculture, Municipality of Tanza.
2 This section draws upon some data and qualitative material presented in Kelly (1998). For other accounts, see: Ramos (ed.) (1991), Ochoa and Descanz (1993), Canlas (1991), and Sermeno (1994).
3 'Land prices in Metro towns, Calabarzon rising', *Philippine Daily Inquirer*, July 1995.
4 This practice has been analysed by Medalla and Centeno (1994).
5 'Cavite has high rate of illegal land conversions, claim farmers groups', *Business World*, 10 June 1991.
6 Of the legal conversions, 254.2 hectares were for residential uses, 77.1 for industrial, 13.1 for institutional, 11.8 for tourism resorts and 11.1 were for unspecified uses. Source: unpublished data on land conversion, Municipal Agrarian Reform Office, Tanza, Cavite, 1995.
7 Unpublished data on farmers in Tanza, 1989 and 1995, Municipal Agricultural Officer, Tanza, Cavite.
8 Administrative Order No. 20, 1992
9 'Gangster politics thrives in Cavite', *Philippine Daily Inquirer*, 2 May 1995.
10 Executive Order No.124 , 1993.
11 Requirements for Conversion from Department of Agrarian Reform, Tanza.
12 Examples of popular protest against development in Cavite include a new dumpsite servicing Manila to be located in Carmona; the Export Processing Zone in Rosario; and various residential and industrial developments in Dasmarinas, most notably against a new industrial estate developed by the Japanese Marubeni Corporation in Langkaan, Dasmarinas, in 1991. The Langkaan controversy lead to the ousting of the Secretary of Agrarian Reform. See "Cavite dumpsite opposed', *Philippine Daily Inquirer*, 16 March 1992; 'Cavite folk protest dump site project', *Manila Chronicle*, 30 April 1993; 'More Cavite farmers lose lands', *Manila Times*, 8 September 1991; 'NDC-Marubeni issue revisited', *Manila Chronicle*, 28 May 1990; 'The Langkaan Syndrome' by John McAndrew, in *Midweek*, 11 April 1990.
13 McAndrew (1990, 1994) and Coronel (1995) record various instances of provincial law enforcement officers or hired 'goons' being involved in tenant evictions.
14 This estimate is derived from a mapping exercise that identified approximately 450 houses, plus a further 100 houses in areas that were not mapped. The estimate of total population is then calculated on the basis of 5 persons per house, an average that holds true for both survey data and government statistics.
15 It should also be noted that the division into primary and secondary household occupations usually reflects the assumptions made according to gender roles by respondents to the survey (both male and female) rather than necessarily prioritizing occupations according to the income that they yield. Thus men are predominant in primary occupations, and women in the secondary income sources.
16 This data represents approximately one-third of the 1995 village population, but it is likely that it cannot be extrapolated in a linear fashion for the entire population. The survey sampling scheme was not random and, as explained earlier, systematically excluded several areas of the village for logistical reasons: Santa Cecilia Subdivision, Retirees I and the 'Squatter Settlement' indicated on map 5. The 'Squatter Settlement' in particular is a recent addition to the village and therefore probably contains a high proportion of migrants. It is

therefore likely that the proportion of migrants in the village's population as a whole is even higher than the estimates provided here.
17 Table 5.7 must, however, be interpreted with the caveats mentioned earlier in mind. The exclusion of the 'Squatter Settlement' together with a number of other houses scattered among rice fields means that the results probably underestimate the absolute numbers engaged in agricultural activities among new migrants to Mulawin.
18 Social housing units elsewhere in Tanza have been bought by the national government and are to be occupied by resettled victims of the Pinatubo eruption and subsequent damage in Central Luzon.
19 One example is the Carissa low-cost housing subdivision in Tanza, build by Villar Homes, a company owned by Congresssman M. B. Villar, now Speaker of the Philippine House of Representatives.
20 *Lahar* is the mixture of ash and water that originates in volcanic deposits. Large areas of farmland in Central Luzon are still being devastated by *lahar* flows from the Pinatubo eruption of 1991 – it is to this image that the farmer is alluding.
21 Department of Agriculture technicians reported that other towns in Cavite have developed similar concentrations on particular crops that have become tied in to a marketing chain with buyers from Manila, for example, *ampalaya* in Dasmarinas and bell peppers in Kawit.

6 RESISTING AND RE-IMAGINING THE GLOBAL

1 Influential works have included *The Lichauco Paper* (Lichauco, 1973), *Nationalist Economics* (Lichauco, 1988), *Dissent and Counter-Consciousness* (Constantino, 1970), and *Neocolonial Identity and Counter Consciousness* (Constantino, 1978).
2 In many Filipino universities, Constantino's historical texts and essays on colonialism are widely used. In particular, Constantino (1975) and Constantino and Constantino (1978).
3 R. Constantino, 1996, syndicated column, 'Outflanking governments', reproduced on the World Wide Web: http://www.mabuhay.com/phil_reporter/...
4 R. Constantino, 1996, syndicated column, 'Globalization hype', reproduced on the World Wide Web: http://www.mabuhay.com/phil_reporter/...
5 Cited by Philippine New Agency, 5 October 1998: PNA10051812.
6 Recent and comprehensive overviews of Philippine civil society movements are provided by Miranda (1997) and in a three-volume collection of essays under the title *Philippine Democracy Agenda* (Diokno, 1997; Wui and Lopez, 1997; Ferrer, 1997).
7 This split emerged in 1992 with the publication of a paper by the founding chairman of the Communist Party of the Philippines, Armando Liwanag. The paper, entitled 'Reaffirm our Basic Principles and Rectify the Errors' sought to reaffirm Marxist–Leninist principles in the face of revisionist tendencies. Circulation of the paper brought to the fore differences that had started to emerge in 1990. The unity of the movement was finally shattered when its chairman, Jose Maria Sison, then in exile in the Netherlands, sent faxes denouncing several individuals within the movement as government agents. The movement became deeply divided between 'reaffirmist' and 'rejectionist' camps. While NGOs leaders in Manila were not necessarily Party members, the left movement as a whole divided into these two camps, hence undermining the unity that had existed for over two decades. For a detailed account, see Rocamora (1994).
8 Extracted from letter of invitation to potential participants.
9 Draft Manifesto of the Peoples' Global Action against Free Trade and the WTO, 1998.
10 These included: Foundation for Independent Analysis and Foundation for Independent Aotearoa (New Zealand); Karnataka State Farmers' Association (India); Movemento sem Terra (Brazil); Movement for the Survival of the Ogoni People (Nigeria); and Mama 86 (Ukraine).
11 Preliminary conference circular, August 1998.

12 See also, 'Antilabor setup in export zones scored', *Today*, 22 February 1995; 'Remulla scored for union busting', *Philippine Daily Inquirer*, 1 May 1995; 'Unions want Remulla probed', *Philippine Daily Inquirer*, 21 March 1995.
13 See 'Cavite dumpsite opposed', *Philippine Daily Inquirer*, 16 March 1992; 'Cavite folk protest dump site project', *Manila Chronicle*, 30 April 1993; 'More Cavite farmers lose lands', *Manila Times*, 8 September 1991; 'NDC-Marubeni issue revisited', *Manila Chronicle*, 28 May 1990; 'The Langkaan Syndrome' by John McAndrew, in *Midweek Magazine*, 11 April 1990.
14 See, for example: Canlas (1991), Ochoa and Descanz (1993), Ramos (ed.) (1991), and Sermeno (1994).
15 A more detailed description of this episode is provided in Kelly (1997).
16 This account of the protest revolving around the power plant is derived from interviews with those involved in the struggle and press accounts.
17 See Calabarzon Watch, 1993. 'Alcala accused of graft'. *Calabarzon Watch Newsletter*, 3(11–12): Sept./Oct., p. 7.
18 'FVR intervention won the battle for Velasco', *Philippine Daily Inquirer*, 12 May 1995.
19 'Ramos takes big swipe at Remulla', *Philippine Daily Inquirer*, 2 May 1995.
20 'Cavite labor problems', *Philippine Daily Inquirer*, 12 April 1995.
21 'Remulla bows to Velasco; Bong in, too', *Philippine Daily Inquirer*, 10 May 1995.
22 'Remulla, Velasco vow no bloodshed', *Philippine Daily Inquirer*, 7 May 1995; 'Remulla bows to Velasco; Bong in, too', *Philippine Daily Inquirer*, 10 May 1995. The election was relatively free from violent incidents, exceptions being the discovery of unexploded home-made bombs at the Tanza National High School and in Dasmarinas, and a shooting in Cavite City. See 'Cavite bombing attempts foiled: voting relatively peaceful in hot spot', *Philippine Daily Inquirer*, 9 May 1995.
23 'FVR intervention won the battle for Velasco', *Philippine Daily Inquirer*, 12 May 1995.
24 The election was not completely trouble-free. The entire province was placed under the security control of the national Commission on Elections which had to deal with various cases if illegal bearing of arms. In addition, on election day a crude bomb was found and successfully defused at Tanza's high school which was serving as a polling station.
25 'Loss due to no union policy: Remulla admits failure', *Philippine Daily Inquirer*, 15 May 1995; see also 'The very best for Cavite', *Philippine Daily Inquirer*, 22 May 1995.
26 In addition, Velasco's tenure as governor was shortlived. By 1997, he had been recruited to a cabinet level post running the Department of Interior and Local Government. He was replaced as governor by his running mate, Bong Revilla, the thirty-year-old son of a prominent senator.

7 CONCLUSION

1 'Manila government "will remain pro-business"', *The Straits Times*, 13 October 1998: 15.

REFERENCES

Agnew, J. and Corbridge, S. 1995. *Mastering Space: Hegemony, Territory and International Political Economy.* London: Routledge.

Agoncillo, A. and Alfonso, A. 1961. *History of the Filipino People.* Quezon City: Ateneo de Manila University Press.

Amin, A. and Thrift, N. 1994. 'Living in the global'. In Amin, A. and Thrift, N. (eds.) *Globalization, Institutions, and Regional Development in Europe.* Oxford: Oxford University Press.

Anderson, B. 1988. 'Cacique democracy in the Philippines: origins and dreams'. *New Left Review,* 169: 3–33.

Appadurai, A. 1990. 'Disjuncture and difference in the global cultural economy'. *Public Culture,* 2: 1–24.

Arcellana, E. 1996. *The Relevance of Recto Today: A Review of Philippine–American and Other Relations.* Quezon City: University of the Philippines.

Balisacan, A. M. 1994. *Poverty, Urbanization and Development Policy: A Philippine Perspective.* Quezon City: University of Philippines Press.

Ball, R. 1997. 'The role of the state in the globalization of labour markets: the case of the Philippines'. *Environment and Planning A,* 29: 1603–28.

Barnes, T. 1996. *Logics of Dislocation: Models, Metaphors and Meanings of Economic Space.* New York: Guilford Press.

Bello, W., Kinley, D. and Elinson, E. 1982. *Development Debacle, the World Bank in the Philippines.* San Francisco: Institute for Food and Development Policy.

Biersteker, T. 1995. 'The "triumph" of liberal economic ideas in the developing world'. In Stallings, B. (ed.) *Global Change, Regional Response: The New International Context of Development.* Cambridge: Cambridge University Press.

Board of Investments. 1998. Unpublished data. Makati City: Board of Investments.

— n.d. *The Philippines: Your Competitive Advantage.* Makati City: Board of Investments.

Bonner, R. 1987. *Waltzing with a Dictator: The Marcoses and the Making of an American Policy.* New York: Times Books.

Boyce, J. K. 1993. *The Philippines: The Political Economy of Growth and Impoverishment in the Marcos Era.* Honolulu: University of Hawaii Press.

Broad, R. 1988. *Unequal Alliance: The World Bank, the International Monetary Fund and the Philippines.* Berkley: University of California Press.

Broad, R. and Cavanagh, J. 1993. *Plundering Paradise: The Struggle for the Environment in the Philippines.* Berkeley: University of California Press.

Canlas, C. 1991. *Calabarzon: The Peasants' Scourge.* Saliksik: Philippine Peasant Institute Research Papers.

REFERENCES

Caoili, M. A. 1988. *The Origins of Metropolitan Manila: A Political and Social Analysis.* Quezon City: New Day.

Castells, M. 1989. *The Informational City.* Oxford: Blackwell

—1996. *The Rise of the Network Society.* Oxford: Blackwell.

—1997. *The Power of Identity.* Oxford: Blackwell.

—1998. *End of Millennium.* Oxford: Blackwell.

Chan, F.-Y. 1999. 'Local politics and labour regulation in Subic Bay, Philippines'. Honours thesis, Southeast Asian Studies Programme, National University of Singapore.

Chant, S. and McIlwaine, C. 1995. *Women of a Lesser Cost: Female Labour, Foreign Exchange and Philippine Development.* East Haven: Pluto Press.

Chua, B.-H. 1995. *Communitarian Ideology and Democracy in Singapore.* London: Routledge.

Constantino, R. 1970. *Dissent and Counterconsciousness.* Quezon City: Malaya Books.

—1975. *The Philippines: A Past Revisited.* Vol. I. Quezon City: The Foundation for Nationalist Studies.

—1978. *Neocolonial Identity and Counter Consciousness: Essays on Cultural Decolonization.* London: Merlin Press.

—1982. *The Miseducation of the Filipino.* Quezon City: The Foundation for Nationalist Studies.

—1985. *The Relevant Recto.* Quezon City: Karrel.

—1991. *A Filipino Vision of Development: Proposals for Survival, Renewal and Transformation.* Quezon City: The Foundation for Nationalist Studies.

Constantino, R. (ed.) 1989. *The Essential Tañada.* Quezon City: Karrel.

Constantino, R. and Constantino, L. R. 1978. *The Philippines: The Continuing Past.* Vol. II. Quezon City: The Foundation for Nationalist Studies.

Corbridge, S., Thrift, N. and Martin, R. 1994. *Money, Power and Space.* Oxford: Blackwell.

Coronel, S. 1995. 'The killing fields of commerce'. In Lacaba, J. F. (ed.) *Boss: Five Case Studies of Local Politics in the Philippines.* Manila: Philippine Centre for Investigative Journalism

Coronel, S. (ed.) 1998. *Pork and Other Perks: Corruption and Governance in the Philippines.* Pasig City: Philippine Centre for Investigative Journalism.

Cullather, N. 1994. *Illusions of Influence: The Political Economy of United States–Philippines Relations, 1942–1960.* Stanford: Stanford University Press.

Cullinane, M. 1971. 'Implementing the "New Order": the structure and supervision of local government during the Taft era'. In Owen, N. G. (ed.) *Compadre Colonialism: Philippine–American Relations, 1898–1946.* Manila: Solidaridad Publishing House.

Cumings, B. 1993. 'Rimspeak; or, The Discourse of the "Pacific Rim"'. In Dirlik, A. (ed.) *What is in a Rim? Critical Perspectives on the Pacific Region Idea.* Boulder: Westview.

Cushner, N. P. 1971. *Spain in the Philippines: From Conquest to Revolution.* Quezon City: Ateneo de Manila University Press.

DAR (Department of Agrarian Reform) 1995. Unpublished data. Quezon City: Department of Agrarian Reform.

DTI (Department of Trade and Industry, Republic of the Philippines) (n.d.). *Framework Plan: Export-led Balanced Agri-Industrialization.* Manila: DTI.

Dicken, P. 1998. *Global Shift: Transforming the World Economy.* London: Paul Chapman Publishing.

Dicken, P. and Thrift, N. 1992. 'The organization of production and the production of organization: why business enterprises matter in the study of geographical industrialization'. *Transactions, Institute of British Geographers,* 17(3): 279–91.

Diokno, M. (ed.) 1997. *Democracy and Citizenship in Filipino Political Culture.* Quezon City: University of the Philippines Third World Studies Centre.

REFERENCES

Doner, R. F. 1991. *Driving a Bargain: Automobile Industrialization and Japanese Firms in Southeast Asia*. Berkeley: University of California Press.
—— 1992. 'Politics and the growth of local capital in Southeast Asia: auto industries in the Philippines and Thailand'. In McVey, R. (ed.) *Southeast Asian Capitalists*. Ithaca: Cornell Southeast Asia Program.
Doronila, A. 1992. *The State, Economic Transformation, and Political Change in the Philippines, 1946–1972*. Oxford: Oxford University Press.
Douglass, M. 1989. 'The environmental sustainability of development: coordination, incentives, and political will in land use planning for the Jakarta Metropolis'. *Third World Planning Review*, 11(2): 211–38.
—— 1991. 'Transnational capital and the social construction of comparative advantage in Southeast Asia'. *Southeast Asian Journal of Social Sciences*, 19(1): 14–43.
—— 1995. 'Global interdependence and urbanization: planning for the Bangkok mega-urban region'. In McGee, T. G. and Robinson, I. M. (eds.) *The Mega-Urban Regions of Southeast Asia*. Vancouver: UBC Press.
Drakakis-Smith, D. 1990. 'Theories of development'. In Dwyer, D. (ed.) *Southeast Asian Development: Geographical Perspectives*. Harlow: Longman.
Dumont, J.-P. 1993. 'The Visayan male Barkada: manly behavior and male identity on a Philippine Island'. *Philippine Studies*, 41: 401–36.
Endriga, J. 1970. 'The Friar Lands settlement: promise and performance'. *Philippine Journal of Public Administration*, 14(4): 397–413.
English, L. J. 1986. *Tagalog–English Dictionary*. Manila: National Book Store Inc.
Eviota, E. 1992. *The Political Economy of Gender: Women and the Sexual Divison of Labour in the Philippines*. London: Zed Books.
Featherstone, M. 1990. 'Global culture: an introduction'. *Theory, Culture and Society*, 7(2–3): 1–14.
—— 1995. *Undoing Culture: Globalization, Postmodernism and Identity*. London: Sage.
Featherstone, M. and Lash, S. 1995. 'Globalization, modernity and the spatialization of theory: an introduction'. In Featherstone, M., Lash, S. and Robertson, R. (eds.) *Global Modernities*. London: Sage.
Feder, E. 1983. *Perverse Development*. Quezon City: The Foundation for Nationalist Studies.
Fegan, B. (1989). 'Accumulation on the basis of an unprofitable crop'. In G. Hart (ed.) *Agrarian Transformations*. Berkeley: University of California Press.
Ferrer, M. (ed.) 1997. *Civil Society Making Civil Society*. Quezon City: University of the Philippines Third World Studies Centre.
Ferreria, A. J., Batalla, C. R. and Magallanes, N. G. 1993. 'Investments and TNCs in the Philippines: towards balanced local and regional development'. *Regional Development Dialogue*, 14(4): 67–93.
Firman, T. 1995. 'The emergence of extended metropolitan regions in Indonesia: Jabotabek and Bandung metropolitan area'. *Review of Urban and Regional Development Studies*, 7: 167–88.
—— 1996. 'Urban development in Bandung metropolitan region'. *Third World Planning Review*, 18(1): 1–21.
Forbes, D. 1986. 'Spatial aspects of Third World multinational corporations' direct investment in Indonesia'. In Taylor, M. and Thrift, N. (eds.) *Multinationals and the Restructuring of the World Economy: The Geography of Multinationals, Vol. 2*. London: Croom Helm.
—— 1997. 'Metropolis and megaurban region in Pacific Asia'. *Tijdschrift voor Economische en Social Geographie*, 88(5): 457–68.

Forbes, D. and Thrift, N. 1987. 'International impacts on the urbanization process in the Asian region: a review'. In Fuchs, R. J., Jones, G. W. and Pernia, E. M. (eds.) 1987. *Urbanization and Urban Policies in Pacific Asia*. Boulder: Westview Press.

Fortune Magazine, 1972. Advertisement, 12 October 1972.

Friedman, J. 1995. 'Global system, globalization and the parameters of modernity'. In Featherstone, M., Lash, S. and Robertson, R. (eds.) *Global Modernities*. London: Sage.

Fuchs, R. J. and Pernia, E. M. 1987. 'External economic forces and national spatial development: Japanese direct investment in Pacific Asia'. In Fuchs, R. J., Jones, G. W. and Pernia, E. M. (eds.) 1987. *Urbanization, Urban Policies in Pacific Asia*. Boulder: Westview Press.

Giddens, A. 1990. *The Consequences of Modernity*. Stanford: Stanford University Press.

— 1991. *Modernity and Self-Identity: Self and Society in the Late Modern Age*. Stanford: Stanford University Press.

Ginsburg, N., Koppel, B. and McGee, T. G. 1991. *The Extended Metropolis: Settlement Transition in Asia*. Honolulu: University of Hawaii Press.

Golay, F. H. 1961. *The Philippines: Public Policy and National Economic Development*. Ithaca: Cornell University Press.

Goodno, J. B. 1991. *The Philippines: Land of Broken Promises*. London: Zed Books.

Gonzalez, J. 1997. 'Political economy of Philippine development: past issues and current reforms'. *Humboldt Journal of Social Relations*, 23(1–2): 91–119.

— 1998. *Philippine Labour Migration: Critical Dimensions of Public Policy*. Singapore: ISEAS.

Granovetter, M. 1985. 'Economic action and social structure: the problem of embeddedness'. *American Journal of Sociology*, 91(3): 481–510.

Greenberg, C. 1994. 'Region based urbanization in Bangkok's extended periphery'. Ph.D. thesis, University of British Columbia, Department of Geography.

Guerrero, S. H., Endencia, D. C. and Bautista, G. M. 1987. *Regional and Socio-Economic Impacts of Export Processing Zones in the Philippines*. Quezon City: Institute of Social Work and Community Development, University of the Philippines, Diliman.

Guttierez, E. 1993. *The Ties That Bind*. Manila: Philippine Centre for Investigative Journalism.

Habito, C. 1993. 'The 1993–1998 MTPDP: Paving the way for Philippines 2000'. *Kasarinlan, A Philippine Quarterly of Third World Studies*, 9(2–3): 7–18

Hannerz, U. 1990. 'Cosmopolitans and locals in world culture'. *Theory, Culture and Society*, 7(2–3): 237–52.

Harvey, D. 1982. *The Limits to Capital*. Oxford: Basil Blackwell.

— 1989. *The Condition of Postmodernity*. Oxford: Blackwell.

— 1990. 'Between space and time: reflection on the geographical imagination'. *Annals, Association of American Geographers*, 80: 418–34.

Hawes, G. 1987. *The Philippine State and the Marcos Regime: The Politics of Export*. Ithaca: Cornell University Press.

Henderson, J., Kuncoro, A. and Nasution, D. 1996. 'The dynamics of Jabotabek development'. *Bulletin of Indonesian Economic Studies*, 32(1): 71–95.

Herod, A. 1997. 'Labour as an agent of globalization and as a global agent'. In Cox, K. (ed.) *Spaces of Globalization: Reasserting the Power of the Local*. New York: Guilford.

Herrin, A. and Pernia, E. 1987. 'Factors influencing the choice of location: local and foreign firms in the Philippines'. *Regional Studies*, 21(6): 531–41.

Hirst, P. and Thompson, G. 1996. *Globalization in Question: The International Economy and the Possibilities of Governance*. Cambridge: Polity Press.

REFERENCES

Hollnsteiner, M. 1963. *The Dynamics of Power in a Philippine Municipality*. Quezon City: The Community Development Research Council, University of the Philippines.

Ibon, 1998. *OCED's Renewed Push for Investment Liberalization*. People's Policy and Advocacy Studies, Special Release, March 1998.

Ileto, R. 1979. *Pasyon and Revolution: Popular Movements in the Philippines*. Quezon City: Ateneo de Manila University Press.

Intal, P. and Llanto, G. 1998. 'Financial reform and development in the Philippines, 1980–1997: imperatives, performance and challenges'. *Philippine Institute for Development Studies, Discussion Paper*, No. 98-02.

Intal, P. and Medalla, E. 1998. 'The East Asian crisis and Philippine sustainable development'. *Philippine Institute for Development Studies, Discussion Paper*, No. 98-04

International Monetary Fund 1996. *International Financial Statistics Yearbook*. New York: IMF.

Islam, R. (ed.) 1987. *Rural Industrialisation and Employment in Asia*. New Delhi: International Labour Organisation.

JICA (Japanese International Cooperation Agency) 1991. *The Masterplan Study on the Project Calabarzon, Final Report*. Manila: Republic of the Philippines, Department of Trade and Industry.

Jessop, R. 1999. 'Some critical reflections on globalization and its illogic(s)'. In Olds, K., Dicken, P., Kelly, P., Kong, L. and Yeung, H. (eds.) 1999. *Globalisation and Asia-Pacific: Contested Territories*. London: Routledge.

Jomo K. S. (ed.) 1997. *Southeast Asia's Misunderstood Miracle: Industrial Policy and Economic Development in Thailand, Malaysia and Indonesia*. Boulder: Westview.

Jurado, G. 1998. 'Global capital: the Philippines in the regional currency crisis. *Public Policy*, 11(1): 16–50.

Kelly, P. F. 1997. 'Globalization, power and the politics of scale in the Philippines'. *Geoforum*, 28: 151–71.

— 1998. 'The politics of urban–rural relations: land use conversion in the Philippines'. *Environment and Urbanization*, 10(1): 35–54.

— 1999a. 'Everyday urbanization: the social dynamics of development in Manila's extended metropolitan region'. *International Journal of Urban and Regional Research*, 23(2): 284–304.

— 1999b. 'The geographies and politics of globalization'. *Progress in Human Geography*, 23(3): 379–400.

— 1999c. 'Rethinking the "local" in labour markets: the consequences of cultural embeddedness in a Philippine growth zone'. *Singapore Journal of Tropical Geography*, 20(1): 56–75.

Kelly, P. F. and Armstrong, W. 1996. 'Villagers and outsiders in cooperation: experiences from development praxis in the Philippines'. *Canadian Journal of Development Studies*, 17(2): 241–59.

Kelly, P. F. and Olds, K. 1999. 'Questions in a crisis: the contested meanings of globalisation in the Asia-Pacific'. In Olds, K., Dicken, P., Kelly, P., Kong, L. and Yeung, H. (eds.) 1999 *Globalization and Asia Pacific: Contested Territories*. London: Routledge.

Kerkvliet, B. J. 1977. *The Huk Rebellion: A Study of Peasant Revolt in the Philippines*. Quezon City: New Day.

— 1990. *Everyday Politics in the Philippines: Class and Status Relations in a Central Luzon Village*. Berkeley: University of California Press.

Koppel, B. 1990. 'Mercantile transformations: understanding the state, global debt and Philippines agriculture'. *Development and Change*, 21: 579–619.

Lamberte, M. B., Manasan, R. G., Llanto, G. M., Villamil, W. M., Tan, E. S., Fajardo, F. C. and Kramer, M. 1993. *Decentralization and Prospects for Regional Growth*. Manila: Philippine Institute for Development Studies.

Laquian, A. 1996. 'Metro Manila: the exceptional city'. Manuscript, Centre for Human Settlements, University of British Columbia.

Larkin, J. 1992. *Sugar and the Origins of Modern Philippine Society*. Berkeley: University of California Press.

Lash, S. and Urry, J. 1994. *Economies of Signs and Space*. London: Sage.

Leaf, M. 1994. 'The suburbanisation of Jakarta: a concurrence of economic and ideology'. *Third World Planning Review*, 16(4): 341–56.

— 1996. 'Building the road for the BMW: culture, vision, and the extended metropolitan region of Jakarta'. *Environment and Planning A*, 28: 1617–35.

Lefebvre, H. 1991. *The Production of Space*. Oxford: Blackwell.

Lichauco, A. 1973. *The Lichauco Paper: Imperialism in the Philippines*. New York: Monthly Review Press.

— 1988. *Nationalist Economics*. Quezon City: Institute for Rural Industrialization.

Lieban, R. W. 1967. *Cebuano Sorcery: Malign Magic in the Philippines*. Berkeley: University of California Press.

Lopez 1966. The Colonial Relationship. In Golay, F. (ed.) *The United States and the Philippines*. New Jersey: Prentice Hall.

Magno, A. 1993. 'A Powerful Icon'. *Kasarinlan: A Philippine Quarterly of Third World Studies*, 9(2/3): 1–2.

McAndrew, J. 1990. 'The Langkaan Syndrome'. *Midweek Magazine*, 11 April.

McAndrew, J. P. 1994. *Urban Usurpation: From Friar Estates to Industrial Estates in a Philippine Hinterland*. Quezon City: Ateneo de Manila University Press.

McCoy, A. W. 1982. 'Introduction: the social history of an archipelago'. In McCoy, A. W. and de Jesus, E. C. (eds.) 1982. *Philippine Social History: Global Trade and Local Transformations*. Quezon City: Ateneo de Manila University Press.

— 1994a. 'An anarchy of families: the historiography of state and family in the Philippines'. In McCoy, A. W. (ed.) 1994. *An Anarchy of Families: State and Family in the Philippines*. Quezon City: Ateneo de Manila University Press.

— 1994b. 'Rent-seeking families and the Philippine state: a history of the Lopez family.' In McCoy, A. W. (ed.) 1994. *An Anarchy of Families: State and Family in the Philippines*. Quezon City: Ateneo de Manila University Press.

McGee, T. G. 1967. *The Southeast Asian City*. London: Bell.

— 1989. 'Urbanisasi or Kotadesasi? Evolving patterns of urbanisation in Asia'. In Costa, F. J. et al. (eds) *Urbanisation in Asia: Spatial Dimensions and Policy Issues*. Honolulu: University of Hawaii Press.

— 1991. Presidential Address. 'Eurocentrism in geography: the case of Asian urbanization'. *The Canadian Geographer*, 35(4): 332–442.

McGee, T. G. and Greenberg, C. 1992. 'The emergence of extended metropolitan regions in ASEAN, 1960–1980, an exploratory outline'. In Pongsapich, A., Howard, M. C. and Amyot, J. (eds.) *Regional Development and Change in Southeast Asia in the 1990s*. Bangkok: Social Research Institute, Chulalongkorn University.

McGee, T. G. and Robinson, I. (eds.) 1995. *The Mega-Urban Regions of Southeast Asia*. Vancouver: UBC Press.

Massey, D. 1984. *Spatial Divisions of Labour: Social Structures and the Geography of Production*. London: Macmillan.

— 1994. *Space, Place and Gender*. Minneapolis: University of Minnesota Press.
May, G. 1987. *A Past Recovered*. Quezon City: New Day.
Medalla, E. M. 1988. Trade and industrial policy beyond 2000: An assessment of the Philippine economy. *Philippine Institute for Development Studies, Discussion Paper*, No. 98-01.
Medalla, F. and Centeno, L. 1994. 'Land use, urbanization and the land conversion issue'. Manuscript, School of Economics, University of the Philippines.
Medhi, Krongkaew. 1996. 'The changing urban system in a fast-growing city and economy: the case of Bangkok and Thailand'. In Lo, F.-C. and Yeung, Y.-M. (eds.) *Emerging Cities in Pacific Asia*. Tokyo: UNU Press.
Medina, I. 1994. *Cavite Before the Revolution*. Quezon City: College of Social Sciences and Philosophy, University of the Philippines, Diliman.
Merrifield, A. 1993. 'Place and space: a Lefebvrian reconciliation'. *Transactions, Institute of British Geographers*, 18: 516–31.
Miranda, F. 1997. *Democratization: Philippine Perspectives*. Quezon City: University of the Philippines Press.
Mulder, N. 1990. 'Philippine textbooks and the national self-image'. *Philippine Studies*, 38: 84–102.
Municipality of Tanza, 1995. *Comprehensive Land Use Plan of the Municipality of Tanza*. Tanza, Cavite: Municipality of Tanza.
Naisbitt, J. 1995. *Megatrends Asia: Eight Asian Megatrends that are Reshaping our World*. New York: Simon and Schuster.
NSO (National Statistics Office) 1990. *Census of Population and Housing, 1990*. Manila: NSO.
— 1993. *Census Facts and Figures*. Manila: NSO.
— 1997. *1995 Census of Population Report No. 2-29 D: Socio-Economic and Demographic Characteristics, Cavite*. Manila: NSO.
NEDA (National Economic and Development Authority) 1995a. *Medium Term Philippine Development Plan, 1993–1998*. Manila: NEDA.
— 1995b. *The Philippines: Back in Business in the Gateway to Asia and the Pacific*. Manila: NEDA.
— 1995c. *Philippine Statistical Yearbook*. Manila: NEDA.
— 1997. *Philippine Statistical Yearbook*. Manila: NEDA.
Nestor, C. 1997. 'Foreign investment and the spatial pattern of growth in Vietnam'. In Dixon, C. and Drakakis-Smith, D. (eds.) *Uneven Development in Southeast Asia*. Aldershot: Avebury.
Ochoa, C. and Descanzo, C. 1993. 'Coverting lands, wrecking lives'. *Philippine Peasant Institute, Briefing Paper*, 11(1), May.
Ofreneo, R. E. 1980. *Capitalism in Philippine Agriculture*. Quezon City: The Foundation for Nationalist Studies.
— 1995. *Philippine Industrialization and Industrial Relations*. Quezon City: University of Philippines, Centre for Integrative Development Studies.
Ohmae, K. 1995a. 'Putting global logic first'. *Harvard Business Review*, Jan.–Feb.: 119–25.
— 1995b. *The End of the Nation State: The Rise of Regional Economies*. New York: The Free Press.
Ong, A. 1987. *Spirits of Resistance and Capitalist Discipline: Factory Women in Malaysia*. Albany: State University of New York Press.
Owen, N. 1971. 'Philippine economic development and American policy: a reappraisal'. In Owen, N. G (ed.) *Compadre Colonialism: Philippine–American Relations, 1898–1946*. Manila: Solidaridad Publishing House.

REFERENCES

Parnwell, M. and Wongsuphsawat, L. 1997. 'Between the global and the local: extended metropolitanisation and industrial location decision making in Thailand'. *Third World Planning Review*, 19(2): 119–38.

Pernia, E. M. 1988. 'Urbanization and spatial development in the Asian and Pacific Region: trends and issues'. *Asia Development Review*, 6(1): 86–105.

Pernia, E. and Israel, R. 1994. 'Spatial development, urbanization and migration patterns in the Philippines'. Paper presented at NEDA conference, Makati, Philippines.

Pernia, E. and Paderanga, C. 1983. 'The spatial and urban dimensions of development'. In Pernia, E. M., Paderanga, C. W. and Hermoso, V. P. (eds.) *The Spatial and Urban Dimensions of Development in the Philippines*. Manila: Philippine Institute for Development Studies.

PEZA (Philippine Economic Zones Authority). 1998. Unpublished data. Manila: Philippine Economic Zones Authority.

Phelan, J. 1959. *Hispanization of the Philippines: Spainish Aims and Filipino Responses, 1565–1700*. Madison: University of Wisconsin Press.

Pinches, M. 1994. 'Modernisation and the quest for modernity: architectural form, squatter settlements and the new society in Manila'. In Askew, M. and Logan, W. S. (eds.) *Cultural Identity and Urban Change in Southeast Asia: Interpretative Essays*. Victoria: Deakin University Press.

Pinches, M. and Lakha, S. (eds.) 1992. *Wage Labor and Social Change*. Quezon City: New Day.

Piven, F. 1995. 'Is it global economics or neo-laissez-faire?' *New Left Review*, 213: 107–14.

Province of Cavite, 1990. *Cavite Provincial Development Plan, 1999–2000*. Trece Martires City: Province of Cavite.

— 1995. Unpublished data. Trece Martires City: Planning Department, Province of Cavite.

— no date. *An Invitation to Invest in Cavite, Philippines*. Trece Martires City: Office of the Provincial Governor.

Putzel, J. 1992. *A Captive Land. The Politics of Agrarian Reform in the Philippines*. London: Catholic Institute for International Relations.

Quigley, K. 1995. 'Environmental organizations and democratic consolidation in Thailand'. *Crossroads*, 9(2): 1–29.

Rafael, V. L. 1988. *Contracting Colonialism: Translation and Christian Conversion in Tagalog Society under Early Spanish Rule*. Quezon City: Ateneo de Manila University Press.

— 1995. 'Introduction: Writing outside: on the question of location'. In Rafael, V. (ed.) *Discrepant Histories: Translocal Essays on Filipino Culture*. Manila: Anvil Publishing.

Ramos, C. (ed.) 1991. *Calabarzon Master Plan: Issues and Implications*. Manila: Ramon Magsaysay Award Foundation.

Reid, A. 1988. *Southeast Asia in the Age of Commerce, 1450–1680. Volume One: The Lands Below the Winds*. New Haven: Yale University Press.

— 1993. *Southeast Asia in the Age of Commerce 1450–1680. Volume Two: Expansion and Crisis*. New Haven: Yale University Press.

Reyes, G. and Paderanga, C. 1983. 'Government policies and spatial development'. In Pernia, E. M., Paderanga, C. W. and Hermoso, V. P. (eds.) *The Spatial and Urban Dimensions of Development in the Philippines*. Manila: Philippine Institute for Development Studies.

Riedinger, J. M. 1995. *Agrarian Reform in the Philippines: Democratic Transitions and Redistributive Reform*. Stanford: Stanford University Press.

Rivera, T. 1994. *Landlords and Capitalists: Class, Family and State in Philippine Manufacturing*. Quezon City: University of the Philippines Press.

REFERENCES

Robertson, R. 1990 . 'Mapping the global condition: globalization as the central concept'. *Theory, Culture and Society*, 7(2–3): 15–30.

—— 1992. *Globalization: Social Theory and Global Culture*. London: Sage.

Rocamora, J. 1994. *Breaking Through: The Struggle Within the Communist Party of the Philippines*. Pasig City: Anvil Publishing.

Rodan, G. 1997. 'Civil society and other possibilities in Southeast Asia'. *Journal of Contemporary Asia*, 27(2): 156–78.

Rodan, G. (ed.) 1996. *Political Oppositions in Industrialising Asia*. London: Routledge.

Ross, H. and Poungsomlee, A. 1995. 'Environmental and social impact of urbanisation in Bangkok'. In Rigg, J. (ed.) *Counting the Costs*. Singapore: Institute of Southeast Asian Studies.

Roth, D. 1982. 'Church lands in the agrarian history of the Tagalog region'. In McCoy, A. W. and de Jesus, E. C. (eds.) *Philippine Social History: Global Trade and Local Transformations*. Quezon City: Ateneo de Manila University Press.

Schirmer, D. and Shalom, D. (eds.) 1987. *The Philippines Reader: A History of Colonialism, Neocolonialism, Dictatorship and Resistance*. Boston: South End Press.

Scott, A. 1990. *Ideology and the New Social Movements*. London: Unwin Hyman.

Scott, W. 1994. *Barangay: Sixteenth Century Philippine Culture and Society*. Manila: Ateneo de Manila University Press.

Seabrook, J. 1996. *In the Cities of the South: Scenes from a Developing World*. New York: Verso.

Sermeno, D. 1994. *Circumventing Agrarian Reform: Cases of Land Conversion*. Pulso (Institute on Church and Social Issues, Manila), Monograph 14, July.

Setchell, C. 1995. 'The growing environmental crisis in the world's mega-cities: the case of Bangkok'. *Third World Planning Review*, 17(1): 1–18.

Sidel, J. 1994. 'Walking in the shadow of the big man: Justiniano and failed dynasty building in Cavite, 1935–1972'. In McCoy, A. (ed.) *An Anarchy of Families: State and Family in the Philippines*. Quezon City: Ateneo de Manila University Press.

—— 1995. 'Coercion, capital, and the post-colonial state: Bossism in the Philippines'. Unpublished Ph.D. thesis, Cornell University.

—— 1997. 'Philippine politics in town, district, and province: Bossism in Cavite and Cebu'. *Journal of Asian Studies*, 56(4): 947–66.

—— 1998. 'The underside of progress: land, labor and violence in two Philippine growth zones, 1985–1995'. *Bulletin of Concerned Asian Scholars*, 30(1): 3–12.

Smith, N. 1992a. 'Contours of a spatialized politics: homeless vehicles and the production of geographical scale, *Social Text*, 33: 54–81.

—— 1992b. 'Geography, difference and the politics of scale'. In Doherty, J., Graham, E. and Malek, M. *Postmodernism and the Social Sciences*. London: Macmillan.

—— 1993. 'Homeless/global: scaling places'. In Bird, J. et al. (eds.) *Mapping the Futures: Local Cultures, Global Change*. London: Routledge.

—— 1996. 'Spaces of vulnerability: the space of flows and the politics of scale'. *Critique of Anthropology*, 16(1): 63–77.

Smith, N and Katz, C. 1993. 'Grounding metaphor: towards a spatialized politics'. In Keith, M. and Pile, S. (eds.) *Place and the Politics of Identity*. London: Routledge.

Soegijoko, B. T. 1996. Jabotabek and Globalisation. In Lo, F.-C. and Yeung, Y.-M. (eds.) *Emerging Cities in Pacific Asia*. Tokyo: UNU Press.

Steinberg, D. J. 1990. *The Philippines: A Singular and a Plural Place*. Boulder: Westview.

Steinberg, D. J. (ed.) 1987. *In Search of Southeast Asia: A Modern History*. Honolulu: University of Hawaii Press.

Sturtevant, D. R. 1976. *Popular Uprisings in the Philippines, 1840–1940*. Ithaca: Cornell University Press.
Swyngedouw, E. 1997. 'Neither global nor local: "glocalization" and the politics of scale'. In Cox, K. (ed.) *Spaces of Globalization: Reasserting the Power of the Local*. London: Guilford.
Takahashi, A. 1970. *Land and Peasants in Central Luzon*. Honolulu: East-West Center Press
Tañada, W. 1993. 'Is there no room for nationalism in the medium-term development plan?' *Kasarinlan, A Philippine Quarterly of Third World Studies*, 9(2–3): 88–94.
Tay, S. 1998. 'Towards a Singaporean civil society'. In Da Cunha, D. and Funston, J. (eds.) *Southeast Asian Affairs, 1998*. Singapore: ISEAS.
Thrift, N. and Olds, K. 1996. 'Refiguring the economic in economic geography'. *Progress in Human Geography*, 20(3): 311–37.
Tigno, J. 1997. 'People empowerment: looking into NGOs, Pos and selected organizations'. In Miranda, F. (ed.) *Democratization: Philippine Perspectives*. Quezon City: University of the Philippines Press.
Tolentino, A. and Ibañez, R. 1983. 'Ancillary firm development in the Philippine automobile industry'. In Odaka, K. (ed.) 1983. *The Motor Vehicle Industry in Asia: A Study of Ancillary Firm Development*. Singapore: Singapore University Press.
Tomlinson, J. 1991. *Cultural Imperialism: A Critical Introduction*. London: Pinter Publishing.
Toye, J. 1993, *Dilemmas of Development – Reflections on the Counter-Revolution in Development: Theory and Policy*. 2nd edn. Oxford: Basil Blackwell.
Turner, M. 1995. 'Subregional economic zones, politics and development: the Philippines involvement in the East ASEAN Growth Area (EAGA)'. *The Pacific Review*, 8(4): 637–48.
Wallerstein, I. 1974. *The Modern World-System*. New York: Academic Press.
Warr, P. 1984. 'Export promotion via industrial enclaves: the Philippines' Bataan export processing zone'. *University of the Philippines School of Economics Discussion Paper*, No. 8407, November 1984.
— 1985. 'Export processing zones in the Philippines'. *ASEAN-Australia Economic Papers*, No. 20.
Waters, M. 1995. *Globalization*. New York: Routledge.
Wickberg, E. 1965. *Chinese in Philippine Life, 1850–1898*. New Haven: Yale University Press.
Wolf, D. 1992. *Factory Daughters: Gender, Household Dynamics, and Rural Industrialization in Java*. Berkeley: University of California Press.
Wolters, W. 1985. *Politics, Patronage and Class Conflict in Central Luzon*. Quezon City: New Day Publishers.
Wui, M. and Lopez, M. G. (eds.) 1997. *State–Civil Society Relations in Policy-Making*. Quezon City: University of the Philippines Third World Studies Centre.
Yeung, H. W. C. 1994. 'Critical reviews of geographical perspectives on business organizations and the organization of production: towards a network approach'. *Progress in Human Geography*, 18(4): 460–90.
Yoshihara, K. 1985. *Philippine Industrialization: Foreign and Domestic Capital*. Quezon City: Ateneo de Manila University Press.

INDEX

Agnew, J. 11
agrarian reform 81, 119, 130, 132–3;
    see also Department of Agrarian
    Reform
agriculture
    attitudes to working in 103, 109, 134–5, 162
    labour process in 84–9
    viability of 138–9
Amaya 151–3, 156, 158, 164
America see United States
American Chamber of Commerce 33
anomie 128
Appadurai, Arjun 7
aquaculture 138
Aquino, Benigno 37
Aquino, Corazon, President 37–9, 64, 81, 146, 161
Asian financial crisis 10, 42–5, 141, 144, 148, 159–60

'balanced' development 94, 112–13
Bangkok 63
barangays 19–20, 24, 133, 156
barkadas 103–5
bataan 33, 78
BAYAN New Patriotic Alliance 148
Bell Trade Act (1946), US 30, 143
Biersteker, T. 12
Board of Investments 33, 39, 52, 54–8, 60, 64, 66
Bocalan, Lino 69
'bossism' 38, 69, 143
Britain 22–3
Broad, R. 11
build-operate-own (BOO) schemes 52
build-operate-transfer (BOT) schemes 39, 52

Bunga 15, 74, 78–113, 114, 126–9, 162
    data gathering in 166–8

Calabarzon zone 41, 65–6, 68, 94
Capipisa 69
capitalism 4–5, 147
Carmona 151
Castells, Manuel 3–5, 8–14 *passim*
Catholicism 24–5, 145
Cavanagh, J. 11
Cavite 1–2, 28, 58–62, 65, 70–71, 73, 78, 94, 151, 160–64
    historical evolution of 68
    local government in 69–70
    promotion of 74–6
    see also Calabarzon zone
Cavite Energy Corporation (CEC) 151
Cavite Export Processing Zone (CEPZ) 1, 33, 55–6, 93, 99–101, 109–12, 114, 119, 122, 124–5, 151, 155
    investment in 128
Cebu 71
centralization 30; see also decentralization
child labour 77, 92–3, 100–102
Chinese in the Philippines 22, 24, 31
civil society organizations 145–50 *passim*
class structure see social structure
clientalism see patron-client relationships
Cojuangco, Eduardo 154
colonialism 17–18, 45–6, 142
    American 16, 18, 26–30, 45
    Spanish 16, 20–26, 45, 68
Confesor, Nieves 77
Constantino, Renato 29, 45, 141–4
Contemplacion, Flor 43
Corbridge, S. 11
Coronel, Sheila 43
corruption, bureaucratic 134, 140, 153

185

criminality 68–9
cronyism 35–6, 43–4
cropping patterns 83–4, 110–11
culture 6–7, 112–13, 163
  of work 102–5
Cumings, B. 41

DAI Young Apparel 77
datus 19, 23–4
Davao City 41
decentralization 1–2, 45, 49, 64, 71, 118
democratic processes 28–9, 45
Department of Agrarian Reform (DAR) 115, 117–18, 134
Department of Environment and Natural Resources 152–3
Department of Labour and Employment 77
Department of Trade and Industry 65
dependent development 141–2
deregulation 1–2, 40, 45, 52
'desakota' regions 60, 62
devaluation of the peso 32, 34, 44–5
devolution 64, 71
Diokno, Jose 31, 143
divisions of labour 89–91, 109–12, 162
Doronila, A. 37
Douglass, M. 41
Dumont, Jean-Paul 103–4

East Asian Growth Area (EAGA) 41
economic zones 41, 57–9
  investments in 52–4
education 29–30, 79, 100
  about colonialism 142
El Nino 44
elites in Filippino society 16, 26–31 *passim*, 38, 45–6, 145, 160–61
embeddedness 6, 14–15, 46, 114, 140, 143, 150, 161–5
employment 55–60, 78
English language 26, 43
environmental concerns 135, 140, 152–3, 162
Estrada, Joseph, President 45, 144–5, 160
exchange controls 31–2
exchange rate of the peso 30
Export Development Act (1994) 40
export-oriented industrialization (EOI) strategy 33, 35, 38, 57, 65, 69, 160–61
export processing zones 1–2, 33, 39, 41, 57, 64–5, 77–8; *see also* Cavite

export of labour *see* overseas workers
export trade 22, 26–38 *passim*

factory work
  as incentive for migration 131
  as point of contact with globalized economy 99
  as seondary source of income 95
  employees favoured for 162–3
  impact on agriculture 94
family relationships 20, 82, 101–2
Featherstone, M. 7
Fegan, Brian 90
fish farming 138
foreign direct investment (FDI) 47, 49–53, 144, 161
  measurement of 52
  sources of 54, 142
  spatial patterns of 56–63
Foreign Investments Act (1991) 39, 42
foreign investors
  incentives for 144
  requirements of 67, 160
Fortune Magazine 34
Foundation for Nationalist Studies 144
Friedman, J. 7

galleon trade 21–2
gender relations and roles 20, 55, 91–2, 100–101, 109–12, 157, 163
Giddens, Anthony 5–6
global scale, privileging of 10–11, 13–14, 45, 164
globalization
  alternatives to 148, 165
  ambivalent response to 158
  definition of 6–7
  discourse of 2, 11–14, 17, 74, 140, 143, 159, 161, 164–5
  geography of (in the Philippines) 63–7
  individuals experience of 162–3
  logics of 4–6
  material processes of 3–14 *passim*, 140
  of production 6
  of resistance to economic processes 147–8
  of social interaction 6
  openness of Philippine economy to 17, 42, 46, 161
  politics of 11–12, 14, 17–18, 71, 159–60
  power geometry of 156–7, 163

resistance to 15, 141, 143, 146–8, 150, 153, 156–8, 164–5
supposed inevitability of 13–14, 143, 159
'glocalization' 11
Golay, F.H. 29
Gordon, Richard 71

Hannerz, U. 7
Hanoi 63
harvesting 105–7, 110, 112, 137
Harvey, David 4–5, 7
Hawes, Gary 36
health and safety at work 76
Herrin, A. 67
hierarchical structures 24–5, 45
history of the Philippines
  pre-colonial 18–20
  see also colonialism
Ho Chi Minh City 63
hybridization 17, 163

Ibon Foundation 146–8
Iglesia Ni Cristo 76
import substitution 31–2, 35
independence, Philippine 30
industrial estates 2, 41, 55, 57, 59, 65, 115, 119, 151
Industrial Security Action Group 78
industrialization 144, 146
informational age, the 5
infrastructure 66–7
Intal, P. 44
integrated area development projects 64
International Monetary Fund (IMF) 16, 32–6, 44, 52
interviews for research 166–8
investment *see* foreign direct investment
irrigation 85–6, 132–7 *passim*
Israel, R. 49, 67

Jakarta 63
Japan
  economic assistance from 65
  occupation of the Philippines 30, 78–9
Jessop, R. 9
'Juan Kaunlaran' 40, 42

Karrel Press Incorporated 144
Kerkvliet, B.J. 91
Kilusang Magbubukid ng Pilipinas 147
Kilusang Mayo Uno (KMU) 146

labour, export of *see* overseas workers
labour markets, local 2, 15, 93–4, 105–7, 112–13
labour mobility *see* migrant workers
labour process
  agricultural 84–9
  regulation of 75–8, 111–12, 151
  see also divisions of labour
labour shortages 149; *see also* harvesting
labour supply, Filipino 42, 70
Laguna 58–9, 65; *see also* Calabarzon zone
land conversion 114–21, 125, 130, 139, 149, 160, 162
  illegal 117
  losers from 137–8, 157
  opposition to 149–51, 155–6
  political involvement in 119
  social and economic factors favouring 132–8, 150
land holding 81–2; *see also* tenancies
land left idle 130, 135, 162
land use 15, 68–9, 115, 118
land values 115, 130
Langkaan Controversy 151
language 25–6, 43
Lash, S. 7
Legazpi, Miguel Lopez de 19, 21, 24
liberalization 1–2, 35, 40, 42, 45, 51, 118, 145–7
  of trade 12–13
Lichauco, Alejandro 142
livestock rearing 89
loans 88
local government role in planning 68–71, 133
location of industry 67
Lopez clan 36
Luzon 23, 49, 57, 64, 126

Macapagal, Diosdado, President 32
McCoy, Alfred 70
McGee, T.G. 60–61
Magno, Alex 40
Malaysia 160
Manila 18–19, 21–3, 34, 47, 49, 56, 63–4, 126
  hinterland of 60
  light rail transit system 39, 67
manufacturing 31, 49, 125
Marcos, Ferdinand, President 16, 32–9 *passim*, 51, 69–71, 81, 119, 145–6, 161

Marcos, Imelda 34, 64
martial law 33, 35, 161
Marubeni Corporation 151
Massey, D. 156, 163
Medalla, E. 44
media influence 135
Medium Term Philippine Development Plan 40, 66, 144
mega-urbanization 47, 49, 56, 62–3
melons, cultivation of 138–9
mestizos 22, 31
migrant workers 49, 125–6, 128, 131
  agricultural 107–9, 112, 137, 163
  see also overseas workers
military alliances 6
military bases 30, 143
military coups 39–40, 145
Mindanao 67
minimum wages 77, 101, 155
missionaries 24–5
Mitra, Ramon 154
modernity and modernization 5–6, 30
monopoly marketing boards 34, 36
Montano family 69–70, 121
motorized hand tractors 84
Mulawin 15, 108, 114–16, 120–25, 128–9, 132–3, 135–8, 149, 162
  data gathering in 166–8
Multilateral Agreement on Investments (MAI) 146

Naisbitt, John 8
nation-states 6
National Food Administration 134
National Statistical Coordination Board 52
nationalism, economic 31, 141–8, 164
neoliberal economc policies 11–13
Nestor 63
New Social Movements (NSM) 150–51
newly-industrializing countries 40
non-governmental organizations (NGOs) 141–2, 145, 148, 152–3, 156, 158, 164–5

occupational structure 96–8, 122–5
Ohmae, Kenichi 8
openness of Philippine economy *see* globalization
Organization for Economic Cooperation and Development (OECD) 146
Osmena family 71

overseas workers, Filipino 1, 34–5, 42–3, 93, 95–9, 102, 112, 127–8, 160; *see also* remittances

Paderanga, C. 64, 67
patron-client relationships 29, 38, 76, 121, 155
people s organizations (POs) 141–2
Pernia, E. 49, 64, 67
peso, the *see* devaluation; exchange rate
Philippine Economic Zones Authority (PEZA) 41, 52–5, 57, 59
'Philippines 2000' 40
Piven, F. 13
political institutions 5
political stability and instability 45, 51, 67
politicization 36, 68
pollution 136
population structure 94–5, 122
postmodernism 150
power, legal and illegal forms of 68–70
power plant development 151–3
property development 130–34
protectionism 35, 142–3
protest movements 148, 152–3
Puyat family 70

questionnaires, use of 167–8

Rafael, Vicente 14, 25, 161
Ramos, Fidel, President 1–2, 39–41, 51–2, 146, 151, 154–5, 161
recruitment to jobs 76–8, 99–100, 162–3
Recto, Claro 31, 143–4
regional economic structure of the Philippines 47
regional identities within the Philippines 18, 20
regional policy and planning 63–6
Reid, A. 19
'relativization of scale' 9–10, 164
religion 20, 25
religious orders 23–4, 68
remittances from overseas workers 43, 90, 95, 127
Remulla, Juanito 70, 76–7, 119, 151–6 *passim*
Revilla, Bong 155–6
rice cultivation 83–90, 111–12, 149, 162
Rivera, T. 38
Robertson, Roland 6–7
Rosario 1–2, 55, 60, 93

sabog seeding technique 84, 87, 105–6, 112
Scott, A. 150
Seabrook, J. 156
secondary occupations 94–7, 123
secondary sources for research 166, 168
shift working 101
Sicar, Gerardo 32
Sidel, John 68–9, 119
Singapore 145
smuggling 68–9
social imaginary 6–7
social networks 128–9
social relations 140, 162
social structures 19–21, 24–5, 28, 45–6, 68, 90–91
social tensions 108–9
Soros, George 10
Southern Tagalog 57–60, 64, 66, 126
'space of flows' 7–12 *passim*, 159–65 *passim*
Spain *see* colonialism, Spanish
Spanish language 25–6
special economic zones *see* economic zones
squatter settlements 115, 125
strikes 76–7
Structural Adjustment Programme (SAP) 36–7
Subic Bay 41, 65, 71
subsistence farming 88–9, 111, 150
'switchers' 14

Tagalog society 20, 23, 25, 129; *see also* Southern Tagalog
Tañada, Lorenzo 143
Tañada, Wigberto 141, 144
Tanza 69, 93–4, 107, 110, 117, 150–53
tariffs 26, 30, 32
technocrats, influence of 32–8 *passim*
technological change 4–5
tenancies, agricultural 79–82, 86, 90–91, 111–12, 119–21, 132–5, 140, 162
Texas Instruments 33

Thailand 145
trade networks 21–2
trade restrictions 21–2, 26–7, 30–31; *see also* tariffs
Trade Union Congress 78
transportation 63, 66, 79, 83, 86
Truman, Harry 31

union organization 77–8, 100, 146, 155, 160
United States
   Philippine relations with 36–7, 39, 143, 146
   trade with 26–7, 30
   see also colonialism, American
urbanization 114–15
'urbanness', feelings of 128
Urry, J. 7

variable geometry 9–10
vegetable growing 89, 110–12, 124
Velasco, Epimaco 154–6
Villar, Manuel 144
Visayans 107–9, 126, 129, 137

Wallerstein, I. 6
water buffalo (*carabao*) 84, 86
water management 135; *see also* irrigation
Weber, Max 91
work, attitudes to 163; *see also* agricultural work; culture of work
World Bank 16, 34–6, 85, 134
world economy 5, 9, 16–17

yields, agricultural 83, 86–7, 106, 112, 135, 139
Yoshihara, K. 22
youth subculture 104, 157

zoning 130, 133, 151–2